Lecture Notes in Mathematics

Edited by A. Dold, B. Eckmann and F. Takens

Subseries: Australian National University, Canberra
Advisers: L.G. Kovács, B.H. Neumann and M.F. Newman

1456

L.G. Kovács (Ed.)

Groups – Canberra 1989

Australian National University Group Theory Program 1989

Springer-Verlag

Berlin Heidelberg New York London
Paris Tokyo Hong Kong Barcelona

Editor

L.G. Kovács
Mathematics IAS, Australian National University
Canberra 2601, Australia

Mathematics Subject Classification (1980): 20-06

ISBN 3-540-53475-X Springer-Verlag Berlin Heidelberg New York
ISBN 0-387-53475-X Springer-Verlag New York Berlin Heidelberg

This work is subject to copyright. All rights are reserved, whether the whole or part of the material
is concerned, specifically the rights of translation, reprinting, re-use of illustrations, recitation,
broadcasting, reproduction on microfilms or in other ways, and storage in data banks. Duplication
of this publication or parts thereof is only permitted under the provisions of the German Copyright
Law of September 9, 1965, in its current version, and a copyright fee must always be paid.
Violations fall under the prosecution act of the German Copyright Law.

© Springer-Verlag Berlin Heidelberg 1990
Printed in Germany

Printing and binding: Druckhaus Beltz, Hemsbach/Bergstr.
2146/3140-543210 – Printed on acid-free paper

This volume contains some of the papers presented (in one case unfortunately only *in absentia*) during the Australian National University Group Theory Program 1989. The focal point of the Program was the Third International Conference on the Theory of Groups and Related Topics; all but three of the papers here were in fact given during that Conference.

Each paper in this volume has been assessed by at least one referee; the editor is deeply indebted to those who have given so freely of their time.

All authors have declared that their paper is in final form and no similar paper has been or is being submitted elsewhere.

INTRODUCTION

The Third International Conference on the Theory of Groups and Related Topics was held at the Australian National University in Canberra from 25 to 29 September, 1989. In many respects it continued the tradition of the earlier conferences held in 1965 and 1973. The total number of participants was 103. While this is less than the 1973 figure, it is pleasing to be able to report that the number of overseas participants was greater, and their geographical distribution wider, than at either of the earlier conferences. There were in fact 54 overseas participants, and they came from 17 countries: Canada, People's Republic of China, France, Federal Republic of Germany, Hungary, Italy, Japan, Republic of Korea, New Zealand, Pakistan, Philippines, Singapore, Thailand, UK, USA, USSR, Zimbabwe.

The conference was more concentrated than its predecessors (one week instead of two). Nevertheless, the talks ranged widely over the whole field of group theory, paying somewhat greater attention than before to cognate areas where group theory finds its applications. There were some 40 talks in all, of which half were invited lectures. The generally high standard of the talks contributed greatly to the success of the conference.

The majority of the visitors were accommodated in Ursula College, which was the centre for the social activities. We are greatly indebted to the College authorities for the friendly, efficient, and tactful way in which they looked after us. Both the initial reception by the Vice-Chancellor of the Australian National University, Professor L. Nichol, and the highly successful dinner were held at the College. Other highlights of the social program were a party hosted by Bernhard and Dorothea Neumann at their home and an excursion to the Tidbinbilla nature reserve.

Financing a conference in which many participants come from distant countries inevitably poses difficulties. Apart from the registration fees, three major sources of support may be identified. First, the conference was officially sponsored by the International Mathematical Union, the Australian Mathematical Society, and the Australian National University. The considerable financial contribution from these institutions at an early stage was vital to forward planning. Second, the conference was held in conjunction with the Australian National University Group Theory Program 1989, whose other activities included miniconferences on group representations, on computation in groups, on soluble groups, and on Burnside questions. The Program, which attracted over 30 overseas visitors as well as a number of Australian mathematicians, greatly enhanced the value of the conference by enabling overseas mathematicians to extend their stay in Australia. Third, the conference is indebted to the many institutions, both in Australia and overseas, who by one means or another provided travel and living expenses for individual participants. The valuable support of the Universities of Melbourne, La Trobe, Sydney, and New South Wales should be specially mentioned.

Many people worked behind the scenes to ensure the success of the conference. My warmest thanks are extended to all of them, particularly to the members of the Organising and Local Organising Committees and to the Proceedings Editor.

Bernhard Neumann was the driving force behind the first two conferences. Indeed, Bernhard has always been the personification of mathematics in action. What better way, then, to celebrate his impending eightieth birthday (on October 15) than by holding a conference? Members of Bernhard's immediate family and many members of his extended family of colleagues and students were present at the conference. Speaker after speaker paid tribute to his contribution to mathematics and his influence on their lives. A photographic display featuring his mathematical career was mounted at the conference dinner and an evening was devoted to reminiscences. We wish Bernhard many more active and productive years.

> G. E. Wall
>
> (Chairman, Organising Committee)

THE ORGANISING COMMITTEE OF THE CONFERENCE

Dr R. A. Bryce	(Treasurer)
Dr P. J. Cossey	(Secretary)
Professor W. Feit	(International Mathematical Union Representative)
Professor T. M. Gagen	
Professor C. F. Miller	
Professor B. H. Neumann	
Dr M. F. Newman	
Professor C. E. Praeger	
Professor G. E. Wall	(Chairman)

THE LOCAL ORGANISING COMMITTEE OF THE CONFERENCE

Dr R. A. Bryce	
Dr P. J. Cossey	
Dr M. F. Newman	(Chairman)
Dr E. A. Ormerod	
Dr E. S. Scott	

PARTICIPANTS AT THE CONFERENCE

S. I. Adian, Steklov Institute, Academy of Science, Moscow, USSR
I. R. Aitchison, University of Melbourne
J. L. Alperin, University of Chicago, USA

G. Baumslag, City College, City University of New York, USA
K. Bencsath, Manhattan College and New York University, USA
A. J. Berrick, National University of Singapore, Singapore
F. R. Beyl, Portland State University, USA
W. Bosma, University of Sydney
M. Brazil, La Trobe University
C. J. B. Brookes, Cambridge University, UK
M. S. Brooks, Canberra College of Advanced Education
R. M. Bryant, University of Manchester Institute of Science and Technology, U.K.
R. A. Bryce, Australian National University
R. G. Burns, York University, Canada
G. Butler, University of Sydney

C. M. Campbell, University of St Andrews, UK
J. Cameron, Queensland University of Technology
J. J. Cannon, University of Sydney
L. Carbone, University of Melbourne
C. Casolo, University of Udine, Italy
D. Chanthorn, Chiang Mai University, Thailand
R. J. Clarke, University of Adelaide
M. D. E. Conder, University of Auckland, New Zealand
S. B. Conlon, University of Sydney
P. J. Cossey, Australian National University
J. A. Covington, University of Melbourne
M. J. Curran, University of Otago, New Zealand

B. A. Davey, La Trobe University

M. Elzanowski, Portland State University, USA
M. Emerton, Monash University

V. Fedri, University of Florence, Italy
R. P. Felix, University of the Philippines, Diliman, Philippines
D. Flannery, University of Sydney
P. P. Förster, Australian National University

T. M. Gagen, University of Sydney
A. M. Gaglione, US Naval Academy, USA
G. Gamble, University of Melbourne
S. P. Glasby, University of Wellington, New Zealand
J. R. J. Groves, University of Melbourne
C. K. Gupta, University of Manitoba, Canada
N. D. Gupta, University of Manitoba, Canada

G. Higman, University of Oxford, UK
R. B. Howlett, University of Sydney
D. C. Hunt, University of New South Wales

G. D. James, Imperial College, University of London, UK
R. K. James, University of New South Wales

G. Karpilovsky, California State University, Chico, USA
O. H. Kegel, University of Freiburg, Germany
P. E. Kenne, Canberra College of Advanced Education
E. I. Khukhro, Mathematics Institute, Academy of Science, Novosibirsk, USSR
A.-C. Kim, Pusan National University, Republic of Korea
L. G. Kovács, Australian National University .

H. Lausch, Monash University
G. I. Lehrer, University of Sydney
P. Longobardi, University of Naples, Italy
P. J. Lorimer, University of Auckland, New Zealand

D. J. McCaughan, University of Otago, New Zealand
R. N. McKenzie, University of California, Berkeley, USA
T. Maeda, Kansai University, Osaka, Japan
A. Mathas, University of Sydney
J. Michel, Ecole Normale Supérieure, France
C. F. Miller, University of Melbourne

A. Nelson, University of Sydney
B. H. Neumann, Australian National University
P. M. Neumann, University of Oxford, UK
W. D. Neumann, Ohio State University, USA
M. F. Newman, Australian National University
W. Nickel, Australian National University
A. C. Niemeyer, Australian National University

S. Oates-Williams, University of Queensland
E. A. Ormerod, Australian National University

P. P. Pálfy, Mathematics Institute, Academy of Science, Budapest, Hungary
D. P. Parrott, University of Adelaide
T.-A. Peng, National University of Singapore, Singapore
C. E. Praeger, University of Western Australia

R. Rapanut, University of Philippines, Baguio, Philippines
A. H. Rhemtulla, University of Alberta, Canada
R. W. Richardson, Australian National University
G. Rosenberger, University of Dortmund, Germany
J. H. Rubinstein, University of Melbourne

E. M. Sanderson, Queen Mary College, University of London, UK
E. A. Scott, Australian National University
D. Segal, University of Oxford, UK

L. Serena, University of Florence, Italy
F. Shaheen, Quaid-e-Azam University, Pakistan
W. Shi, Southwest China Teachers University, Peoples Republic of China
V. Shpilrain, Moscow State University, USSR
M. W. Short, Australian National University
H. S. Sim, Australian National University
N. F. Smythe, Australian National University
L. H. Soicher, Queen Mary College, University of London, UK
B. Srinivasan, University of Illinois at Chicago, USA
A. G. R. Stewart, University of Zimbabwe, Zimbabwe
R. H. Street, Macquarie University
G. A. Swarup, University of Melbourne

M. C. Tamburini, University of Brescia, Italy
D. E. Taylor, University of Sydney
J. G. Thompson, University of Cambridge, UK

G. E. Wall, University of Sydney
M. A. Ward, Australian National University
M. J. Wicks, National University of Singapore, Singapore
J. Wiegold, College of Cardiff, University of Wales, UK
J. S. Wilson, University of Cambridge, UK

VISITORS TO OTHER PARTS OF THE PROGRAM

B. C. Brewster, State University of New York at Binghamton, USA
P. Donovan, University of New South Wales
M. J. Edeson, Canberra College of Advanced Education
G. Havas, University of Queensland
I. M. Isaacs, University of Madison-Wisconsin, USA
O. K. Kwon, University of Sydney
B. D. McKay, Australian National University
E. A. O'Brien, Marquette University, USA
S. J. Pride, University of Glasgow, UK
G. R. Robinson, University of Manchester Institute of Science and Technology, UK
L. Rylands, University of Sydney
T. A. Springer, University of Utrecht, The Netherlands
R. Stöhr, Karl Weierstraß Institute for Mathematics, Berlin, Germany
G. A. Willis, Australian National University
C. Zworestine, University of Sydney

CONTENTS

SURVEYS

RESEARCH PAPERS

A LIE APPROACH TO FINITE GROUPS

J. L. Alperin

A couple of years ago I received a postal card from Bernhard Neumann bearing the message, "Happy one-half birthday." I replied, thanking him, recalling that I had been in Canberra a few years before on the occasion of his three-quarters birthday, and I let him know that I had just accepted an invitation to speak at the celebration of his four-fifths birthday. Now I would like to add that I am looking forward to his first birthday, in the year 2009.

Group theory is a subject that comes in many different variations: finite groups, infinite groups, Lie groups, transformation groups, algebraic groups and so on. It cuts across algebra, analysis and geometry. Nevertheless, all the different types of groups are closely related, either directly or by analogy. The two oldest kinds of groups, finite groups (where the subject began) and Lie groups are particularly close. We wish to add to that relationship here.

1. STRUCTURE OF LIE TYPE GROUPS

The Lie type groups, the classical groups and the other analogs of Lie groups, are at the heart of finite group theory. Much of their structure is entirely similar with the structure of Lie groups. We shall examine some of the most basic aspects of the structure of Lie type groups from this point of view. For the sake of simplicity of exposition, we shall restrict ourselves to the case of the general linear group $G = GL(n, q)$ over a finite field with $q = p^e$ elements; as is often the case, nothing will be lost in doing this.

The Borel subgroups are the most common and useful of the subgroups. These are the group B of upper triangular matrices and its conjugates. The parabolic subgroups are the conjugates of the subgroups which contain B and these are easily described. Each parabolic containing B is a group of upper block-triangular matrices, that is, the group of all non-singular matrices which, for a fixed sequence of positive integers $k_1, \ldots, k_r$ which sum to n, have square matrices of sizes $k_1, \ldots, k_r$ along the main diagonal and are zero on the remaining entries below this diagonal. Each of these parabolic subgroups is consequently an extension of a p-group (the block-triangular

Supported in part by the Australian National University and grants from the NSA and NSF.

matrices with identity matrices for the square matrices) by a direct product of general linear groups of dimensions $k_1, \ldots, k_r$.

Having examined a bit of the subgroup structure, let us now turn to the geometry associated with the general linear group. If V is the natural n-dimensional module for $GL(n, q)$ then the parabolic subgroups are the stabilizers of flags of subspaces of V, that is, of strictly ordered sequences of non-zero proper subspaces of V. In particular, B is the stabilizer of a complete flag, that is a sequence $V_1, V_2, \ldots, V_{n-1}$, where V_i has dimension i. It is a truly great idea of Tits' that the uniform way to study the different geometries of the Lie and Lie type groups, for example, unitary, symplectic or orthogonal geometries, is to introduce a simplicial complex, that is, a triangulated space, called the building, on which the group acts. In the case of the general linear group the description is quite simple. The vertices of this space, that is, the zero-dimensional simplices, are the non-zero proper subspaces of V. Two vertices are joined by an edge, a one-simplex, if one of the subspaces is properly contained in the other. Three vertices are joined to form a triangle, a two-simplex, if one of the subspaces is strictly contained in a second which in turn is properly contained in the third, that is, the three subspaces form a flag. We now have an immediate and close connection between the subgroup structure we have discussed and the geometry: the parabolic subgroups are the stabilizers of the simplices in the building!

Having discussed subgroup structure and geometry, let us now turn to representation theory. One of the common types of structures that arise here are the Hecke algebras. If P is a parabolic subgroup of G, then the coset space G/P is a G-set and we can form a vector space over a field with this set as basis, so that we get a module for G. The endomorphism algebra of this module is known as a Hecke algebra and these algebras play a central role in the representation theory of G. Notice that we have, at once, a connection with the subgroup structure. The representation theory of groups of Lie type like G, over an algebraically closed field k of characteristic p, where $q = p^e$, is extremely analogous with the finite-dimensional representation theory of Lie groups. In particular, there is a well-developed theory of weights [3]. If S is a simple kG-module and U is the group of upper triangular matrices with ones on the main diagonal (the p-subgroup in the description of the way a parabolic subgroup is an extension in the case of B) then the subspace of S of elements left fixed by U is one-dimensional and the stabilizer of this subspace is a parabolic subgroup containing B (one of the upper block-triangular groups). In this way we have attached a one-dimensional module for a parabolic subgroup to each simple kG-module. It is then true that this defines a one-to-one correspondence (modulo isomorphism and conjugacy, of course) between the simple modules and the one-dimensional representations, the weights, of the parabolic subgroups. Again, the representation theory is closely related to subgroup structure.

Finally, in this survey of some of the most basic ideas that arise in studying Lie type groups, we turn to a result which ties together the subgroup structure, the geometry and the representation theory. Let C_0 be the kG-module which has the 0-simplices, the vertices, of the building as a basis, let C_1 be the kG-module similarly constructed for the 1-simplices, the edges, and so on up to C_{n-2}, the module corresponding to the $n-2$ simplices, that is the complete flags on V. Since the building is a simplicial complex there are boundary maps which are kG-homomorphisms of each C_i to C_{i-1}, for i positive. Moreover, we can let C_{-1} be the one-dimensional trivial module for G, with the empty set as a basis element, and define a boundary map, the augmentation, from C_0 to C_{-1} which sends each vertex to the empty set. In this way we can define a sequence of modules and maps, the augmented chain complex of the building with coefficients in k, as follows:

$$C_{n-2} \to C_{n-3} \to \cdots \to C_1 \to C_0 \to C_{-1} \to 0$$

The image of each boundary map is contained in the kernel of the next one and the successive quotients are the (reduced) homology groups of the building with coefficients in k, a sequence of kG-modules. The theorem of Solomon-Tits [6] states that all these homology groups are zero, with the one exception of H_{n-2} and this is a simple module, which is also projective at the same time, and is known as the Steinberg module, one of the most basic of all the modules. In this way, we have an example of how three different aspects of the structure of groups of Lie type, subgroup structure, geometry and representation theory, are all very closely related.

2. LIE STRUCTURE OF FINITE GROUPS

It has long been observed, in the classification work on simple groups and in the representation theory of arbitrary finite groups, that there are subgroups that arise in the work that appear analogous with the parabolic subgroups of Lie type groups. In fact, there are many aspects of Lie type groups that appear in the study of general finite groups. Our thesis here is that these are not an unrelated set of interesting and provocative accidents, but evidence of an important unifying principle which should be taken very seriously and which is easy to enunciate:

If G is an arbitrary finite group and p is any prime divisor of its order, then there exist interesting and important analogs of all the aspects of the structure of Lie type groups whose natural characteristic is p.

In particular, to take a random example, say, let G be the symmetric group Σ_9 and $p = 3$: then there should be analogs of parabolic subgroups, weights and so on. We shall now illustrate this principle for the ideas we have just discussed in the preceding

section. We begin by tabulating the results of this part of the paper where we shall
discuss analogs for all these ideas. On the left are the Lie type concepts and on the
right the analogous ideas for arbitrary finite groups.

Structure of Lie type groups	Lie structure of finite groups
Borel subgroups	Sylow normalizers
Parabolic subgroups	p-parabolic subgroups
Buildings	Brown complexes
Hecke algebras	Centralizer rings
Weights	Weights
Steinberg modules	Steinberg complexes

The Borel subgroup B is the normalizer of the subgroup U which is, in turn, a
Sylow p-subgroup of $GL(n,q)$. The normalizer of a Sylow p-subgroup is the obvious
analog for an arbitrary group and it is certainly clear that such subgroups play a central
role in group theory, even appearing in a critical way in the most elementary uses of
the Sylow theorems which are taught in first courses in algebra. A subgroup L of an
arbitarary group G is called a p-parabolic if L is the normalizer in G of the largest
normal p-subgroup $O_p(L)$ of L; in the Lie type case this property has long been known
to characterize parabolic subgroups, by a theorem and Borel and Tits. The p-parabolic
subgroups occur in many places; for example, the normalizer of a defect group of a
p-block is a p-parabolic subgroup.

Turning to geometry, there is indeed a very good analog of the building for our
arbitrary group G, the Brown complex due to K. Brown [2]. Consider the poset (partially ordered set) $S_p(G)$ of all non-identity p-subgroups of G. We define a simplicial
complex $|S_p(G)|$ by letting the vertices be these non-identity p-subgroups, joining two
vertices if one of the two subgroups is properly contined in the other, forming a triangle
from three vertices that are a strictly linearly ordered set of three subgroups, and so on
in the same way as we constructed the building but using a different poset, not the poset
of non-zero proper subspaces of a vector space. The connection with buildings is quite
direct: if G is of Lie type and characteristic p then the Brown complex and the building
are of the same homotopy type [4]. (In fact, there is a closer connection: Let $|B_p(G)|$
be the Bouc complex which is the simplicial complex formed from the poset $B_p(G)$ of
non-identity p-subgroups Q of G such that $N(Q)$ is a p-parabolic subgroup of G with
$Q = O_p(N(Q))$; $|B_p(G)|$ is then also of the same homotopy type as the Brown complex
and, if G is of Lie type and characteristic p, is homeomorphic with the building being,
in fact, isomorphic with the first barycentric subdivision of the building). The Brown
complex and related complexes are now of great interest to a number of group theorists
and it is clear that they are basic objects of study in the theory of finite groups.

Turning to representation theory, the analogs of Hecke algebras are well-known and precede the introduction of Hecke algebras; they are the centralizer rings of Schur. If G acts transitively on a set X then the endomorphism ring of the module, formed from linear combinations of elements of X, is a centralizer ring and is a useful object in the study of permutation groups and in applications to representation theory. The tie between these two concepts is so close that it is now customary to call all centralizer rings by the name Hecke algebras! It is also possible to define the concept of weight for our arbitrary group G; the concept does not just apply in the specialized Lie situation. A pair (Q, S) is called (see [1]) a weight of G if Q is a p-subgroup of G and S is a simple kG-module for $N(Q)$ which is also projective when regarded as a module for $N(Q)/Q$ (which may be done as Q must act trivially on S since S is simple). (Of course, changes due to conjugacy or isomorphism are not regarded as changing the weight.) In the Lie type case these weights correspond naturally with the Lie weights; in fact, the Steinberg module plays a role in this connection. Moreover, there is a conjectured relation with simple kG-modules: the number of simple kG-modules equals the number of weights. This conjecture and its refinements have attracted a great deal of attention recently and a solution would represent a real breakthrough in representation theory. A reformulation due to Knörr and Robinson [5] is particularly suggestive since it involves use of the Brown complex and puts the above weight conjecture and the Alperin-McKay conjecture, on characters of height zero in a block, in a very similar form.

Finally, we wish to again tie up subgroup structure, geometry and representation theory in a way analogous to the Solomon-Tits theorem on the Steinberg module. Here, the decisive result is one of Webb's [7]. We can form an augmented chain complex for the Brown complex just as we did for the building. Webb's result is that this chain complex of kG-modules is the direct sum of two complexes, one consisting of just projective modules and the other complex having the property that it is a long exact sequence which is also split (which is called contractible and is analogous to the geometric idea). One may assume that the complex consisting of projective modules has no summand which is a contractible complex and it is then unique up to isomorphism, by the Krull-Schmidt theorem applied to complexes. If G is of Lie type then the projective complex just described consists of a complex all of whose modules are zero, with just one exception, the Steinberg module which appears as the component in one dimension. For this reason, it is entirely appropriate to call the complex of projective modules appearing in Webb's theorem by the name of the Steinberg complex. Webb has also shown how these results may be used in an extremely effective way to calculate cohomology of modules, a problem that hitherto seemed much more difficult, so that these ideas are not only interesting as analogs but they are very useful and basic.

3. Homology of the Brown complex

We now wish to apply this Lie principle to show how it leads to new mathematics. There are a number of topics which would illustrate this but we shall discuss the question of the nature of the homology of the Brown complex. This should certainly be interesting if the Brown complex is a good analog of the building. Webb's theorem also suggests this topic since the structure of the Steinberg complex is quite unknown. We shall restrict ourselves to reduced homology with the integers Z as coefficients and write $H_n(S_p(G))$ for $H_n(|S_p(G)|, Z)$, so we are really studying the augmented chain complex for the Brown complex with Z as coefficients in place of k. It would be just as easy to just deal with k as coefficients but we shall use the coefficients that usually appear in topology.

A group with a normal p-subgroup has all the homology of its Brown complex zero; the complex is contractible and the reverse implication is conjectured [4]. In the case that G is of Lie type and of characteristic p the connection between the building and the Brown complex again yields that the homology of the Brown complex is non-zero in exactly one dimension. A theorem of Quillen [4] also gives the same result for a solvable group G with abelian Sylow p-subgroups and with $O_p(G)$ the identity. These results suggest that there is a paucity of non-zero homology for the Brown complex, but nothing of the sort is the case as we shall now illustrate with a couple of theorems.

THEOREM 1. *If the generalized Fitting subgroup $F^*(G)$ is a p'-group and E is a maximal elementary abelian p-subgroup of G and of order p^e, then $H_{e-1}(S_p(G)) \neq 0$.*

(The assumption on $F^*(G)$ is equivalent with the existence of a normal p'-subgroup of G which contains its own centralizer.)

THEOREM 2. *If $p > 3$ and E is a maximal elementary abelian p-subgroup of the symmetric group Σ_n and of order p^e, then $H_{e-1}(S_p(\Sigma_n)) \neq 0$.*

The proofs have a number of ideas in common. One is the use of the Quillen complex $|A_p(G)|$ formed from the poset $A_p(G)$ of non-identity elementary abelian p-subgroups of G which is also of the same homotopy type as the Brown complex and so has the same homology. Another is the use of a complex which appears to be quite unrelated. For any group G and a prime p, we form a complex $|C_p(G)|$ on the set $C_p(G)$ of subgroups of G of order p. The vertices are these subgroups, two are joined by an edge if they commute elementwise, three form a triangle if they again commute elementwise and so on. The 1-skeleton of this complex, the graph consisting of the vertices and edges, is a well-studied graph and occurs in work on the classification of simple groups and in the construction of the Fischer sporadic simple groups. A general result is as follows:

THEOREM 3. *The commuting complex and the Brown complex are of the same homotopy type.*

This result depends on the Quillen complex, use of the so-called maximal cover of a poset (which also shows that the complex, whose vertices are the Sylow p-subgroups of G and where simplices are formed from sets of Sylow subgroups which have a non-identity intersection, is of the same homotopy type as the Brown complex) and other uses of what is called the nerve of a cover. The idea is to use all the techniques based on covers, the simplest of which is the Meyer-Vietoris sequence.

To give a little idea of the proof we shall consider the second theorem. To prove that homology is nonzero, it is necessary, in particular, to show that there exist cycles, that is, chains with zero boundaries; we shall restrict ourselves to exhibiting some spheres that arise in studying the commuting complex of the symmetric groups and which are a critical, but easy step of the argument. Consider a (Young) subgroup $\Sigma_p \times \Sigma_p \times \cdots \times \Sigma_p$ of a symmetric group. Assuming that $p > 3$ there exist distinct Sylow p-subgroup R_1 and T_1 in the first factor and these are two non-commuting subgroups of order p. Similarly, we may choose two Sylow p-subgroups of each of the succeeding factors. The subcomplex of the commuting complex given by R_1 and T_1 consists of two disconnected points, a 0-sphere. If we use in addition the Sylow p-subgroups of the second factor we get a graph which is a square, and so is a 1-sphere, as follows:

If we now use R_3 and T_3 then we will get an octahedron, the suspension of the above 1-sphere and itself a 2-sphere; this may be visualized by thinking of T_3 as lying above the plane of the preceding diagram, connected to each of the four vertices with four triangles also filled in, while R_3 may be thought of lying below the diagram and similarly connected. Continued use of more factors leads to spheres of higher dimension inside the commuting complex. It is, in fact, possible to go back and use the Quillen complex directly, and so eliminate the use of the commuting complex, in the proof of the first two theorems but the ideas are then much more obscure. In any case, it should be apparent now that the homology of the Brown complex promises to be a rich topic.

4. Other topics

We do not wish to close without at least the mention of a number of other topics that fit in with our Lie principle. First, any conjugacy class of elements of order p in a group G defines a subcomplex of the commuting complex; in particular, any class has a well-defined fundamental group as well as all sorts of homology attached to it. These invariants seem to be of interest, at least in some cases. S. Bouc has been studying the case of $p = 2$ and the class of transpositions in a symmetric group. He has proved that the homology need not be free abelian and may have torsion. Even more, he has proved that the fundamental group of the transpositions in the symmetric group Σ_7 is of order three! Bouc suggested that this must be related to the existence of the triple cover of the alternating group on seven letters and the author and G. Glauberman have shown this is just the case by proving a theorem relating covering groups and fundamental groups of conjugacy classes.

Webb's theorem suggest a very general question: How can we describe the structure of a chain complex of a simplicial complex on which a group G is acting? His own result, in its strongest form, suggests a direction, since he studies complexes in which the fixed point set of each non-identity p-subgroup is contractible. The author has proved a very general result which gives, as a special corollary, necessary and sufficient conditions under which the conclusions of Webb's theorem hold. The idea of the proof is simply to carry over to complexes the theory of modular representations, in particular the fundamental Green correspondence.

Finally, we close mentioning what we consider a very basic and general problem, one which has received a lot of attention over the years, which is far from a complete answer and which now seems to be much more tractable with new ideas. This is the question of carrying over to Hecke algebras all of the rich theory of modular representations, so that theory can then be viewed as the special case of such a general theory applied to the case of a group acting transitively on itself, as the Hecke algebra is then the group algebra.

In any case, we hope that we have convinced our readers that it is a good idea to put a little more Lie in their groups!

References

[1] J. L. Alperin, 'Weights for finite groups', in *The Arcata conference on representations of finite groups*, ed. by Paul Fong, Proc. Symposia in Pure Math. vol. 47 Part 1, pp. 369–379 (Amer. Math. Soc., Providence, 1987).

[2] K. Brown, 'Euler characteristic of groups: the p-fractional part', *Invent. Math.* **29** (1975), 1–5.

[3] C. W. Curtis, 'Modular representations of finite groups with split (B, N)-pairs', in *Seminar on algebraic groups and related finite groups*, ed. by A. Borel et al., Lecture Notes in Math. **131**, pp. 57–95 (Springer-Verlag, Berlin, 1970).

[4] D. Quillen, 'Homotopy properties of the poset of nontrivial p-subgroups of a group', *Advances Math.* **28** (1978), 101–128.

[5] R. Knörr and G. Robinson, 'Some remarks on a conjecture of Alperin', *J. London Math. Soc.* (2) **39** (1989), 48–60.

[6] Louis Solomon, 'The Steinberg character of a finite group with *BN*-pair', in *Theory of Finite Groups (Symposium, Harvard Univ.,* 1968), pp. 213–221 (Benjamin, New York, 1969).

[7] P. Webb, 'A split exact sequence for Mackey functors' (preprint).

University of Chicago
CHICAGO IL 60637
USA

ON FINITE BASES FOR LAWS OF TRIANGULAR MATRICES

A. N. Krasil'nikov and A. L. Shmel'kin

In our short survey we shall consider the finite basis problem for laws of groups and Lie algebras of triangular matrices and for laws of representations of groups and Lie algebras by triangular matrices. All terminology and basic facts relating to the laws of groups, Lie algebras, and their representations, can be found in [1–4].

1. Laws of Triangular Matrix Groups
and of Group Representations by Triangular Matrices

B. H. Neumann asked in [5] whether the laws of each group have a finite basis. A. Ju. Ol'shanskii [6], S. I. Adyan [7], and M. R. Vaughan-Lee [8] proved that there exist groups whose laws are not finitely based. However, the existence problem for matrix groups with the same property is not yet solved. It follows easily from [9–11] that any group of 2×2 matrices over a field has a finite basis for its laws. The following theorem implies the existence of a finite basis for the laws of any subgroup of the group $T_n(R)$ of all invertible triangular $n \times n$ matrices over any associative and commutative ring R with 1.

THEOREM 1 [12]. *Any nilpotent-by-abelian group has a finite basis for its laws.*

The existence of a finite basis for the laws of each (nilpotent of class at most 2)-by-abelian group was proved by R. M. Bryant and M. F. Newman [13]. The same property for the laws of nilpotent-by-(abelian of finite exponent) groups was proved in [14].

Let R be an associative and commutative ring with 1. Let F be the free group freely generated by x_1, x_2, ... and RF its group algebra over R. Suppose that G is a group, M is a module over R, and $\varrho : G \to \mathrm{Aut}\,(M)$ is a representation of G on M. We say (see [3]) that an element $f(x_1, \ldots, x_n)$ of the group algebra RF is a law of ϱ if for all g_1, ..., g_n in G we have $f(\varrho(g_1), \ldots, \varrho(g_n)) = 0$.

The results of [6–8] imply the existence of infinite dimensional representations of groups over any field such that their laws are not finitely based. On the other hand, from the results of S. M. Vovsi and Nguen Khung Shon [15] it follows that the laws of any representation of a group by matrices over any finite field have a finite basis. It is not known whether each matrix representation of a group over an infinite field must have a finite basis for its laws, but the following theorem implies the existence of a finite basis for the laws of any triangular matrix representation of a group.

THEOREM 2 [12]. *Let ϱ be a group representation over some Noetherian associative and commutative ring R with 1. Suppose that ϱ satisfies the law*

$$(*) \qquad (x_1 x_2 - x_2 x_1)(x_3 x_4 - x_4 x_3) \cdots (x_{2n-1} x_{2n} - x_{2n} x_{2n-1})$$

for some integer n. Then ϱ has a finite basis for its laws.

In the case of a group representation over a field of characteristic zero, Theorem 2 was proved in [16]; for an outline of the proof in that case, see [17].

2. LAWS OF LIE ALGEBRAS OF TRIANGULAR MATRICES AND OF REPRESENTATIONS OF LIE ALGEBRAS BY TRIANGULAR MATRICES

If R is a finite field or a field of characteristic zero, and if $t_n(R)$ is the Lie algebra of all $n \times n$ triangular matrices over R, then each subalgebra of $t_n(R)$ has a finite basis for its laws. This is a consequence of the following theorems.

THEOREM 3 [18]. *Any nilpotent-by-abelian Lie algebra over a field of characteristic 0 has a finite basis for its laws.*

THEOREM 4 (Ju. A. Bahturin and A. Ju. Ol'shanskii [19]). *Any finite dimensional Lie algebra over a finite field has a finite basis for its laws.*

In the case of an infinite field of characteristic $p > 0$ the situation is more complicated. Let H be the Lie algebra of triangular 3×3 matrices over some infinite field of characteristic 2 spanned by the matrices E_{12}, E_{13}, E_{22}, E_{23} where E_{ij} denotes the matrix with i, j entry 1 and all other entries 0. M. R. Vaughan-Lee proved in [20] that the laws of H are not finitely based. In [21], V. S. Drensky constructed a Lie algebra $H(p)$ of $(p+2) \times (p+2)$ triangular matrices over an infinite field of characteristic $p > 0$ whose laws are not finitely based. On the other hand, the following theorem implies the existence of a finite basis for the laws of any Lie algebra of triangular $n \times n$ matrices over a field of characteristic p, if $n \leqslant p$.

THEOREM 5 [22]. *Over any field of characteristic $p > 0$ any (nilpotent of class at most $(p-1)$)-by-abelian Lie algebra has a finite basis for its laws.*

The existence of a finite basis for the laws of any (nilpotent of class at most 2)-by-abelian Lie algebra over a field of characteristic not 2 was proved by R. M. Bryant and M. R. Vaughan-Lee in [23].

Let L be the free Lie algebra freely generated by x_1, x_2, ... over a field R, and let A be the free associative algebra over R freely generated by the same set, so that $L \subset A$. Suppose that H is a Lie algebra over R, M is a vector space over R, and $\varrho \colon H \to \mathfrak{gl}(M)$ is a representation of H on M. We say (see [4] or [17]) that an element $f(x_1, \ldots, x_n)$ of the free associative algebra A is a law of ϱ if for all h_1, ..., h_n in H we have $f(\varrho(h_1), \ldots, \varrho(h_n)) = 0$.

The following theorem implies that, over a field of characteristic 0 or over a finite field, any representation of a Lie algebra by triangular matrices has a finite basis for its laws.

THEOREM 6 [17]. *Let ϱ be a representation of a Lie algebra over a field of characteristic zero or over a finite field. Suppose that ϱ satisfies the law* (*) *for some integer n. Then the laws of ϱ have a finite basis.*

Over an infinite field of characteristic $p > 0$ there are Lie algebras with triangular matrix representations whose laws are not finitely based.

PROPOSITION 7 [24]. *The adjoint representations of the Lie algebras H and $H(p)$ mentioned above do not have finite bases for their laws. These representations are triangular matrix representations of Lie algebras.*

On the other hand, the following theorem implies the existence of a finite basis for the laws of any representation of a Lie algebra by triangular $n \times n$ matrices over such a field if $n \leqslant p$.

THEOREM 8 [24]. *Let ϱ be a Lie algebra representation over some field of characteristic $p > 0$. Suppose that ϱ satisfies the law* (*) *for some integer $n \leqslant p$. Then the laws of ϱ are finitely based.*

REFERENCES

[1] Hanna Neumann, *Varieties of groups*, Ergebnisse der Mathematik und ihrer Grenzgebiete **37** (Springer-Verlag, Berlin, 1967).

[2] Ju. A. Bahturin, *Lectures on Lie algebras*, Studien zur Algebra und ihrer Anwendungen **4** (Akademie-Verlag, Berlin, 1978).

[3] B. I. Plotkin, 'Varieties of group representations', *Usp. Mat. Nauk.* **32** (1977), 3-68 (in Russian). English translation: *Russian Math. Surveys* **32** (1977) no. 5, 1–72.

[4] Ju. P. Razmyslov, 'Finite basing of the identities of a matrix algebra of second order over a field of characteristic zero', *Algebra i Logika* **12** (1973), 83–113 (in Russian). English translation: *Algebra and Logic* **12** (1973), 47–63 (1974).

[5] B. H. Neumann, 'Identical relations in groups, I', *Math. Ann.* **114** (1937), 506–525.

[6] A. Ju. Ol'shankskii, 'On the problem of a finite basis of identitites in groups', *Izv. Akad. Nauk SSSR Ser. Mat.* **34** (1970), 376–384 (in Russian). English translation: *Math. USSR—Izv.* **4** (1970), 381–389 (1971).

[7] S. I. Adyan, 'Infinite irreducible systems of group identities', *Dokl. Akad. Nauk SSSR* **190** (1970), 499–501 (in Russian). English translation: *Soviet Math. Dokl.* **11** (1970), 113–115.

[8] M. R. Vaughan-Lee, 'Uncountably many varieties of groups', *Bull. London Math. Soc.* **2** (1970), 280–286.

[9] V. P. Platonov, 'Linear groups with identical relations', *Dokl. Akad. Nauk SSSR* **11** (1967), 581–582 (in Russian).

[10] D. E. Cohen, 'On the laws of a metabelian variety', *J. Algebra* **5** (1967), 267–273.

[11] John Cossey, 'Laws in nilpotent-by-finite groups', *Proc. Amer. Math. Soc* **19** (1968), 685–688.

[12] A. N. Krasil'nikov, 'On the finiteness of bases of laws of nilpotent-by-abelian groups' (in Russian, to appear).

[13] R. M. Bryant and M. F. Newman, 'Some finitely based varieties of groups', *J. London Math. Soc.* (3) **28** (1974), 237–252.

[14] A. N. Krasil'nikov and A. L. Shmel'kin, 'On the Specht property and basis rank of some products of varieties of groups', *Algebra i Logika* **20** (1981), 546–554 (in Russian). English translation: *Algebra and Logic* **20** (1981), 357–363 (1982).

[15] S. M. Vovsi and Nguen Khung Shon, 'Identities of almost stable representations of groups', *Mat. Sb.* **132 (174)** (1987), 578–591 (in Russian). English translation: *Math. USSR—Sb.* **60** (1988), 569–581.

[16] A. N. Krasil'nikov, 'On the laws of triangular matrix representations of groups', *Trudy Moskovsk. Mat. Obshch.* **52** (1989), 229–245 (in Russian).

[17] A. N. Krasil'nikov and A. L. Shmel'kin, 'On the laws of finite dimensional representations of solvable Lie algebras and groups', in *Algebra—Some current trends*, ed. by L. L. Avramov and K. B. Tchakerian, Lecture Notes in Math. **1352**, pp. 114–129 (Springer-Verlag, Berlin, 1988).

[18] A. N. Krasil'nikov, 'The finite basis property for certain varieties of Lie algebras', *Vestnik Moskov. Univ. Ser. I. Mat. Mekh.* (1982), no. 2, 34–38 (in Russian). English translation: *Moscow Univ. Math. Bull.* **37** (1982), no. 2, 44–48.

[19] Ju. A. Bahturin and A. Ju. Ol'shanskii, 'Identical relations in finite Lie rings', *Mat. Sb.* **96** (1975), 543–559 (in Russian).

[20] M. R. Vaughan-Lee, 'Varieties of Lie algebras', *Quart. J. Math. Oxford Ser.* (2) **21** (1970), 297–308.

[21] V. S. Drenski, 'On identities in Lie algebras', *Algebra i Logika* **13** (1974), 265–290 (in Russian). English translation: *Algebra and Logic* **13** (1974), 150–165 (1975).

[22] A. N. Krasil'nikov, 'On the laws of nilpotent-by-abelian Lie algebras over a field of finite characteristic' (in Russian, to appear).

[23] R. M. Bryant and M. R. Vaughan-Lee, 'Soluble varieties of Lie algebras', *Quart. J. Math. Oxford Ser.* (2) **23** (1972), 107–112.

[24] A. N. Krasil'nikov, 'On the finiteness of bases of laws of finite dimensional representations of solvable Lie algebras, II' (in Russian, to appear).

A. N. Krasil'nikov
Department of Mathematics
Moscow Pedagogical Institute
Krasnoprudnaya ul. 14.
Moscow 107140
U S S R

A. L. Shmel'kin
Department of Mathematics and Mechanics
Moscow State University
Leninskie Gory
Moscow 119899
U S S R

GROUP REPRESENTATIONS, GEOMETRY AND TOPOLOGY

To Ingse, 1958–88

G. I. LEHRER

1. INTRODUCTION

In 1965 Graham Higman wrote in the Proceedings of the first of these conferences [H]: "It is one of my deeper mathematical convictions that the theory of representations of the general linear groups ... needs to be rewritten every generation or so, in the idiom of the day ... ". The purpose of the present note is not to carry this through completely, but to discuss "the idiom of the day" in the context of representations of GL_n and more generally of reductive groups over finite fields. In particular, our main theme will be the introduction of geometric and topological methods into this theory, which has led to beautiful and sometimes not fully understood interrelationships between groups, geometry and topology.

In § 2 we introduce two broad geometric-topological principles and in § 3 give several examples of their application, especially to the (ordinary and modular) representation theory of finite Lie groups. The remainder of this survey will be devoted to showing how the computation of certain important character values (the 'Green functions') of the finite Lie groups reduces to questions in the cohomology of certain projective complex varieties. We also give a summary of the current state of knowledge on this subject, together with some conjectures and a collection of problems which relate generally to the interconnections between the subjects of our title (see § 7 below).

2. GENERAL PRINCIPLES

The context in which this discussion will take place is that of the following two 'principles', which will be embodied in various contexts below. Because of the diversity of the applications, we state these principles very imprecisely and make precise statements in the specific applications. Suppose g is an endomorphism of the space X; for example, g may be an element of a permutation group or of a group of diffeomorphisms, or g may be an endomorphism of an algebraic variety. We suppose that X has a 'cohomology theory' $H^i(X)$, i.e., that there is a sequence of functors H^i $(i \in \mathbf{Z})$ from a category containing X to vector spaces such that $H^i(X) = 0$ for almost all i.

A. THE LEFSCHETZ PRINCIPLE

We have (in appropriate circumstances)

$$(2.1) \qquad \sum_{i \geqslant 0} (-1)^i \operatorname{tr}(g, H^i(X)) = \chi_E(X^g)$$

where χ_E denotes the Euler characteristic of a space, i.e.,

$$\chi_E(Y) = \sum_{i \geqslant 0} (-1)^i \dim H^i(Y).$$

Note that if X^g is a finite set of points then $\chi_E(X^g) = \#(X^g)$, the cardinality of the fixed-point set of g.

The most primitive example of (2.1) is that of a group G acting as a permutation group on a finite set X, with no additional structure. In this case $H^0(X)$ is just the linearization of the permutation action, and (2.1) expresses the well-known fact that the trace of a group element g on $H^0(X)$ is the number of fixed points of g on X. More generally, (2.1) applies to group actions on finite simplicial complexes (cf. [CL]), to Frobenius endomorphisms of algebraic varieties [D], and there are more sophisticated variations available for intersection cohomology and sheaves of local coefficients on X. The validity of the Lefschetz principle requires that the vector spaces $H^i(X)$ are over fields of characteristic zero, unlike the next principle, some of whose most interesting applications are in positive characteristics.

B. THE HOPF PRINCIPLE

Here we assume that the cohomology spaces $H^i(X)$ are obtained by taking the cohomology of some cochain complex

$$\cdots \longrightarrow C^{i-1}(X) \xrightarrow{\ \partial^i\ } C^i(X) \xrightarrow{\ \partial^{i+1}\ } C^{i+1} \longrightarrow \cdots$$

(where $\partial^{i+1} \partial^i = 0$ for each i). Then we have

$$(2.2) \qquad \sum_{i \geqslant 0} (-1)^i \operatorname{tr}(g, C^i(X)) = \sum_{i \geqslant 0} (-1)^i \operatorname{tr}(g, H^i(X)).$$

In its most elementary form, (2.2) is the classical Hopf trace formula, while more generally it might be thought of as an equation in an appropriate Grothendieck ring of G-modules.

3. EXAMPLES

3.1. THE WEIL CONJECTURES

Let $X(q)$ be the set of solutions $(x_1, x_2, x_3) \in \mathsf{F}_q^3$ (where F_q is the finite field of q elements) of the equation

$$(3.1.1) \qquad x_1^2 + x_2^2 + x_3^2 = 1.$$

Denote by $N(q)$ the cardinality of $X(q)$. The number $N(q)$ may be computed explicitly for q a prime by summing the Legendre symbols $\left(\frac{ax^2 + bx + c}{q}\right)$ over $x \in \mathsf{F}_q$. The result is

$$(3.1.2) \qquad N(q) = \begin{cases} q^2 + q & \text{if } q \equiv 1 \mod 4, \\ q^2 - q & \text{if } q \equiv 3 \mod 4. \end{cases}$$

However, this problem (and similar much more difficult ones) may be put into the context of Principle A above as follows. Let K be the algebraic closure of F_q and denote by $X(K)$ the set of solutions of (3.1.1) in K^3. Define $F \colon X(K) \to X(K)$ by $F(x_1, x_2, x_3) = (x_1^q, x_2^q, x_3^q)$ (this is the q-Frobenius map on $X(K)$). Then $X(q) = X(K)^F$, and the Weil conjectures (now theorems) assert that the computation of $\#(X(K)^F)$ may be carried out using the Lefschetz principle.

(3.1.3) THEOREM (Weil, Grothendieck, Deligne, M. Artin; see [D] for a precise statement). *Let Y be a nonsingular irreducible projective variety defined over F_q. Then there exists a cohomology theory $H^i(Y)$ (ℓ-adic cohomology with compact supports) satisfying (among other things)*

(i) $\sum_{p \geqslant 0} (-1)^p \operatorname{tr}(F, H^p(Y)) = \#(Y^F)$ *(Lefschetz principle);*

(ii) *the eigenvalues of F on $H^i(Y)$ are of the form $\varepsilon q^{i/2}$, where ε is a root of unity;*

(iii) *('Comparison theorem') If $Y(\mathsf{C})$ is the 'corresponding' complex variety, then $\dim H_c^i(Y(\mathsf{C})) = \dim H^i(Y)$, where $H_c^i(Y(\mathsf{C}))$ denotes cohomology with compact supports of a topological space, with complex coefficients.*

Relating this to our example above, we have the following.

(3.1.4) If $X(\mathsf{C}) = \left\{ (x_1, x_2, x_3) \in \mathsf{C}^3 \mid x_1^2 + x_2^2 + x_3^2 = 1 \right\}$, then

$$\dim_{\mathsf{C}} H_c^i(X(\mathsf{C})) = \begin{cases} 1 & \text{if } i = 2 \text{ or } 4, \\ 0 & \text{otherwise.} \end{cases}$$

Together with (3.1.3), this shows that $N(q) = \varepsilon_1 q + \varepsilon_2 q^2$ where ε_1, ε_2 are roots of unity, which may be determined using methods of Deligne [D].

3.2. THE TITS BUILDING

In the context of Principle A, take $X = \mathcal{T}$, the Tits building of a finite group G of Lie type (or, more generally, of a group with a BN-pair). This is a simplicial object whose simplexes are parabolic subgroups of G and whose vertices are the maximal parabolic subgroups of G.

It is well-known that $\mathcal{T}$ has the homotopy type of a bouquet of spheres of dimension $\ell - 1$, where ℓ is the (semisimple) rank of G. For a purely algebraic proof of this, which relates $\mathcal{T}$ to the Coxeter complex of the Weyl group of G, see [CL2]. Included in [CL2] is a proof of the following fact.

(3.2.1) *The representation of G on $H^{\ell-1}(\mathcal{T})$ is irreducible.*

This representation is known as the Steinberg representation of G (written St_G). In view of the vanishing theorem for cohomology its character may be computed using Principle A (see [CLT]).

(3.2.2) THEOREM ([CLT], Theorem 9.2). *Let G be a connected reductive F_q-group. Let $x \in G(q)$. Then*

$$\operatorname{tr}(x, St_G) = \begin{cases} 0 & \text{if } x \text{ is not semisimple} \\ (-1)^{\sigma(G)+\sigma(Z)}|Z_u| & \text{if } x \text{ is semisimple} \end{cases}$$

where Z_u is a maximal unipotent subgroup of $Z_G(x)^0(q)$ and $\sigma(H)$ is the F_q-rank of any F_q-group H.

The proof proceeds by identifying the fixed-point set $\mathcal{B}(G)^x$ explicitly, where $\mathcal{B}(G)$ is a topological realisation of a suspension of $\mathcal{T}$.

3.3. THE STEINBERG REPRESENTATION

Principle B is illustrated in the above context as follows. The cohomology of $\mathcal{T}$ is computed from the chain complex corresponding to the given simplicial subdivision. By the vanishing theorem for the cohomology of $\mathcal{T}$, the right hand side of (2.2) reduces essentially to one term. This leads to the following.

(3.3.1) *With notation as in (3.2),*

$$St_{G(q)} = \sum_{B \subseteq P} (-1)^{n(P)} \operatorname{Ind}_{P(q)}^{G(q)}(1)$$

where B is a fixed Borel subgroup of G and $n(P)$ denotes the semisimple rank of the parabolic subgroup P.

This formula in the Grothendieck ring of G was first proved by Curtis.

3.4. A CHARACTER FORMULA

Continuing the theme of (3.2), and using the notation of Theorem (3.2.2), it was shown in [Le3] that if x is a regular semisimple element of $\mathcal{B}(G)$, then $\mathcal{B}(G)^x$ (the fixed point set of x) is a sphere, whose dimension depends on the 'type' of x (which is described by a conjugacy class in the Weyl group for G split). Using principles A and B, this yields a formula for $\operatorname{Ind}_{P(q)}^{G(q)}(1)(x)$, where P is a parabolic $\mathbf{F}_q$-subgroup of G.

3.5. THE DELIGNE-LUSZTIG THEORY

If G is a reductive $\mathbf{F}_q$-group as in (3.2) and $F\colon G \to G$ is the corresponding Frobenius endomorphism, Deligne and Lusztig [DL] have defined virtual representations $R_T^G(\theta)$ of G^F for any F-stable maximal torus T and character θ of T^F (see §4 below). These are defined as the θ-isotypic components of $\sum_{i=0}^{\infty} H_c^i(X_T, \overline{\mathbf{Q}}_\ell)$, where X_T is a certain variety corresponding to T which is acted on by $G^F \times T^F$, and H_c^i is ℓ-adic cohomology with compact supports.

The crux of the work in [DL] is the computation of the characters of $R_T^G(\theta)$ using Principle A in the form (3.1.3).

3.6. WEBB'S THEOREM

Let G be any finite group and let $\mathcal{P}$ be the poset of non-trivial p-subgroups of G (p a prime). Let X be the corresponding simplicial object (the simplexes of X are the chains in $\mathcal{P}$). This is referred to by Alperin as the 'Brown complex' of G and has been studied by Quillen in [Qu]. One of Quillen's results is that if G is a Lie type group as in (3.2) above, then X is homotopy equivalent to $\mathcal{T}$, the Tits building of G. This has led to a great deal of significant work by Webb [PW] and others on virtual analogues for an arbitrary finite group of the Steinberg representation, which is known in the Lie case to be a factor of any projective G-module in characteristic p. In particular if $C^i(X)$ is the cochain complex of X with coefficients in a field of characteristic p, the following is a consequence of a theorem of Webb.

(3.6.1) *We have*

$$\sum_{i \geqslant 0}(-1)^i C^i(X) = \sum_{i \geqslant 0}(-1)^i D^i(X)$$

where the $D^i(X)$ are projective G-modules.

This equation in the Grothendieck ring of G is an instance of Principle B.

3.7. HYPERPLANE COMPLEMENTS

Let W be a finite Coxeter group and suppose M_W is the (complexified) complement of its reflecting hyperplanes. Then W acts on M_W and hence on its cohomology. This action has been studied in [OS], [LSo], [Le1], [Le2] and many results are available along the lines of both principles A and B.

4. Representation theory of finite reductive groups

For the remainder of this work, we shall discuss some aspects of the impact of geometric and topological methods on the characteristic zero representation theory of the finite groups G^F, where G is a reductive F_q-group and $F \colon G \to G$ is the corresponding Frobenius endomorphism.

(4.1) EXAMPLE. If $G = GL_n(\overline{\mathsf{F}}_q)$ with $\overline{\mathsf{F}}_q$ the algebraic closure of F_q and if $F((a_{ij})) = (a_{ij}^q)$, then $G^F = \{\, (a_{ij}) \mid a_{ij} = a_{ij}^q \,\} = GL_n(\mathsf{F}_q)$.

In (3.5) above mention was made of the virtual characters $R_T^G(\theta)$ of G^F (note that we use the same notation for the virtual module and its character). The character formula for $R_T^G(\theta)$ referred to above reduces the problem of computing all of the characters $R_T^G(\theta)$ on G^F to that of evaluating $R_T^G(1)$ on G_{uni}^F, where G_{uni}^F denotes the set of unipotent element of G^F. This problem is a very important part of the character theory of G^F and we shall show in the next two sections how it essentially reduces to problems in the cohomology of certain complex projective varieties.

(4.2) PROBLEM. *Find the value of $R_T^G(1)$ on each element $u \in G_{\mathrm{uni}}^F$.*

Now the characters $R_T^G(1)$ are parametrized by G^F-conjugacy classes of maximal F-stable tori T of G. To simplify the exposition below, we shall assume that G is F-split (i.e., that F acts trivially on the Weyl group W of G). This assumption is not necessary for most of the results, but reduces the technicalities involved in their statement.

(4.3) LEMMA. *With the above notation, the G^F-conjugacy classes of F-stable maximal tori of G are in bijective correspondence with the conjugacy classes in the Weyl group W of G.*

PROOF. Take an F-split maximal torus T_0 in G; then $W = N_G(T_0)/T_0$ and $F(T_0) = T_0$. Any maximal torus T of G is of the form $T = gT_0g^{-1}$ (some $g \in G$). We have $F(T) = T$ if and only if $g^{-1}F(g) \in N_G(T_0)$. Now the conjugacy class of $g^{-1}F(g)$ in $W = N_G(T_0)/T_0$ is determined by T: for if $T = g_1 T_0 g_1^{-1} = g_2 T_0 g_2^{-1}$ and $F(T) = T$, then $g_1^{-1}g_2 \in N_G(T_0)$ and since G is F-split, F acts trivially on $N_G(T_0)/T_0$ whence $F(g_1^{-1}g_2)g_2^{-1}g_1 \in T_0$. Using the fact that $g_1^{-1}F(g_2) \in N_G(T_0)$, a short computation shows that $[F(g_2^{-1})g_1]F(g_1^{-1})g_1[g_1^{-1}F(g_2)]g_2^{-1}F(g_2) \in T_0$, i.e., that $g_1^{-1}F(g_1)$ and $g_2^{-1}F(g_2)$ lie in the same conjugacy class of W (modulo T_0).

By Lang's theorem ([St]), $g \mapsto g^{-1}F(g)$ is surjective on G, whence every element of W occurs. Finally, $g_1^{-1}F(g_1) = g_2^{-1}F(g_2)$ if and only if $g_1 g_2^{-1} \in G^F$. Hence the G^F-conjugacy class of the F-stable maximal torus $T = gT_0g^{-1} \leqslant G$ is determined by the conjugacy class in W of $g^{-1}F(g)$ (mod T_0). $\qquad\qquad\Box$

(4.4) EXAMPLE. In example (4.2) take T_0 to be the group of diagonal elements. The G^F-conjugacy classes of F-stable maximal tori correspond to conjugacy classes in the symmetric group S_n, and so to partitions $\lambda = (\lambda_1, \lambda_2, \ldots, \lambda_p)$ of n. If T_λ corresponds to λ, then $|T_\lambda| = (q^{\lambda_1} - 1)(q^{\lambda_2} - 1) \cdots (q^{\lambda_p} - 1)$ and T_λ is a direct product of cyclic groups of order $q^{\lambda_i} - 1$ $(i = 1, 2, \ldots, p)$.

In view of (4.3) and the fact that the characters $R_T^G(1)$ depend only on the G^F-conjugacy class of T, for $u \in G_{\mathrm{uni}}^F$ we write $R_T^G(1)(u) = Q_w(u)$, where $w \in W$ is in the conjugacy class of W which corresponds to T.

(4.5) DEFINITION. For $w \in W$, the function Q_w on G_{uni}^F defined above is called the Green function of G^F corresponding to w.

The name comes from Green's pioneering work [**JAG**] on the finite general linear groups.

We now turn to the consideration of the case $w = 1$.

The conjugacy class of tori corresponding to $w = 1$ is represented by T_0 (in the notation of (4.3)) which lies in an F-stable Borel subgroup B of G (e.g., in the case of GL_n, B may be taken as the group of upper triangular matrices and T_0 as the group of diagonal matrices). The Deligne-Lusztig construction simplifies in this case to

$$(4.6.1) \qquad\qquad R_{T_0}^G(1) = \mathrm{Ind}_{B^F}^{G^F}(1).$$

It follows that the Green function Q_1 is given by

$$(4.6.2) \qquad\qquad Q_1(u) = \mathrm{Ind}_{B^F}^{G^F}(1)(u).$$

Now

$$\mathrm{Ind}_{B^F}^{G^F}(1)(u) = \#\{ \text{ cosets } gB^F \in G^F/B^F \mid ugB^F = gB^F \} = \#(G^F/B^F)_u$$

where $(G^F/B^F)_u$ denotes the set of fixed points of u on G^F/B^F. Further,

$$(G^F/B^F)_u = (G/B)_u^F$$

since B is connected (this requires an application of Lang's theorem [**St**]). If we write $\mathcal{B} = G/B$ for the flag variety of G and $\mathcal{B}_u$ for the subvariety of $\mathcal{B}$ fixed by u, then it follows from (3.1.3) that

$$(4.6.3) \qquad \begin{aligned} Q_1(u) &= \#(\mathcal{B}_u^F) = \sum_{p \geq 0} (-1)^p \, \mathrm{tr}(F, H_c^p(\mathcal{B}_u, \overline{\mathbf{Q}}_\ell)) \\ &= \sum_{i \geq 0} \dim H_c^{2i}(\mathcal{B}_u, \overline{\mathbf{Q}}_\ell) q^i. \end{aligned}$$

Thus in the case $w = 1$ the computation of $Q_w(u)$ reduces to a geometric one. In the next section, we show that this is true in general.

5. Geometric Interpretation

(5.1) EXAMPLE. We continue with the example of (4.1) and (4.4), where $G = GL_n$, the Weyl group $W = S_n$, and both the conjugacy classes in W and the unipotent classes are parametrized by partitions $\lambda = (\lambda_1, \lambda_2, \ldots, \lambda_p)$ $(\lambda_1 \geqslant \lambda_2 \geqslant \cdots \geqslant \lambda_p)$ of n. In the former case the λ_i are the cycle lengths of $w \in W$ and in the latter the λ_i are the sizes of the Jordan blocks of n. The following tables of values of $Q_w(u)$ may be found in [JAG]:

$$GL_3$$

$u \backslash w$	(1^3)	$(2,1)$	(3)
(1^3)	$q^3 + 2q^2 + 1$	$1 - q^3$	$q^3 - q^2 - q + 1$
$(2,1)$	$2q + 1$	1	$1 - q$
(3)	1	1	1

$$GL_4$$

$u \backslash w$	(1^4)	(21^2)	(2^2)	(31)	(4)
(1^4)	$q^6 + 3q^5 + 5q^4 + 6q^3 + 5q^2 + 3q + 1$	$1 + q + q^2 - q^4 - q^5 - q^6$	$q^6 - q^5 + q^4 - 2q^3 + q^2 - q + 1$	$1 - q^2 - q^4 + q^6$	$1 - q - q^2 + q^4 + q^5 - q^6$
(21^2)	$3q^3 + 5q^2 + 3q + 1$	$-q^3 + q^2 + q + 1$	$1 - q + q^2 - q^3$	$1 - q^2$	$1 - q - q^2 + q^3$
(2^2)	$2q^2 + 3q + 1$	$q + 1$	$1 - q + 2q^2$	$1 - q^2$	$1 - q$
(31)	$3q + 1$	$q + 1$	$1 - q$	1	$1 - q$
(4)	1	1	1	1	1

In these examples, the following features are apparent.

(5.1.1) The values $Q_w(u)$ are all polynomials in q.

(5.1.2) The leading coefficients of the $Q_1(u)$ (i.e., the coefficients of the highest power of q which occurs) are the degrees of the irreducible representations of W. In fact if u corresponds to the partition λ, the corresponding leading coefficient is the degree of χ_λ (in the usual Frobenius notation, cf. [Mac]).

(5.1.3) Take u to be subregular, i.e., of type $(2,1)$ in the case of GL_3 and of type $(3,1)$ in the case of GL_4. Then $Q_w(u) = 1 + \rho(w)q$, where ρ is the reflection character of W regarded as a Euclidean reflection group.

(5.1.4) The leading coefficients of $Q_w(1)$, i.e., the coefficient of q^3 in type GL_3 and the coefficient of q^6 in type GL_4, are $\varepsilon(w)$, where ε is the sign character of W.

(5.1.5) For each u except u regular, the linear part of $Q_w(u)$ is equal to $1 + q\rho(w)$ where ρ is as in (5.1.3) above.

From the empirical evidence above it seems clear that to understand the values $Q_w(u)$, one should fix u and allow w to vary. Thus we change our notation (cf. [LS]).

(5.2) DEFINITION. For $w \in W$ and $u \in G^F$ write $Q_u(w) = Q_w(u)$.

Thus for each $u \in G^F$ we have a class function Q_u defined on W, with values in $\mathbf{Z}[q]$.

To understand these functions geometrically we introduce the group

$$C(u) = Z_G(u)/Z_G(u)^0$$

of components of the centralizer of u in G; this $C(u)$ classifies the G^F classes into which the G-conjugacy class of u splits (see, e.g., [SSt]), and we denote by u_c a representative in G^F of the class corresponding to $c \in C(u)$. An element x of G^F is called split (or 'distinguished') if F acts trivially on $C(x)$ and on the set of irreducible components of $\mathcal{B}_x$. It is a result of Shoji and Spaltenstein that each F-stable conjugacy class of unipotent elements of G contains a split element of G^F, except in type E_8. Clearly $C(u)$ acts on $\mathcal{B}_u$ and hence on $H_c^i(\mathcal{B}_u, \overline{\mathbf{Q}}_\ell)$.

The geometric interpretation of the Green function comes from

(5.3) THEOREM (Kazhdan, Springer, Lusztig [Ka], [Sp2], [L2]). *In the above notation, let* u *be a split unipotent element of* G^F.
(i) *There is an action of* W *on* $H_c^*(\mathcal{B}_u, \overline{\mathbf{Q}}_\ell)$ *such that for* q *sufficiently large (but in any characteristic) we have for* $c \in C(u)$, $w \in W$,

$$Q_w(u_c) = \sum_{i \geq 0} \operatorname{tr}(w.c, H_c^{2i}(\mathcal{B}_u, \overline{\mathbf{Q}}_\ell)) q^i \in \mathbf{Z}[q].$$

(ii) *If* p *(the characteristic) is sufficiently large, the nilpotent classes in* $\mathcal{G} = \operatorname{Lie} G(\mathbf{C})$ *have the same parametrisation as the unipotent classes in* G. *Let* A *be a nilpotent element of* $\mathcal{G}$ *corresponding to* u. *There is an action of* W *on* $H^*(\mathcal{B}_A)$ *(here* H^* *denotes ordinary complex cohomology) and* $H^*(\mathcal{B}_A) \cong H_c^*(\mathcal{B}_u, \overline{\mathbf{Q}}_\ell)$ *as complex (or* $\overline{\mathbf{Q}}_\ell$*)* W-modules. *In particular, we have (for* u *split, as above)*

$$Q_w(u) = \sum_{i \geq 0} \operatorname{tr}(w, H^{2i}(\mathcal{B}_A)) q^i.$$

Here $\mathcal{B}_A = \{ gB \in G(\mathbf{C})/B(\mathbf{C}) \mid A \in \operatorname{ad} g \operatorname{Lie} B \}$.

The result (i) was known until recently only for sufficiently large p and q. The generalization to arbitrary characteristic is due to Lusztig [L2].

The W-action on $\mathcal{B}_A$ above is the 'Springer action' as defined by Lusztig in [L3]. The action arises not from an action of W on $\mathcal{B}_A$ which is transferred to cohomology functorially in the usual way, but from an action of W on a covering of the dense open subset $\mathcal{G}_{rs}$ of regular semisimple elements of $\mathcal{G}$, which we shall now outline.

Let $\widetilde{\mathcal{G}} = \{(X, gB) \in \mathcal{G} \times \mathcal{B} \mid X \in adg\,\mathrm{Lie}\,B\}$. The projection $\pi \colon \widetilde{\mathcal{G}} \to \mathcal{G}$ is the well-known 'Grothendieck resolution' (cf. Brieskorn [Br2]), while the second projection is a locally trivial fibration with fibre $\mathrm{Lie}\,B$, so that $H^*(\widetilde{\mathcal{G}}) \cong H^*(\mathcal{B})$. Note that if $A \in \mathcal{G}$ then $\pi^{-1}(A) = \mathcal{B}_A$. If $\mathcal{G}_{rs}$ denotes the set of regular semisimple elements of $\mathcal{G}$, write $\widetilde{\mathcal{G}}_{rs} = \pi^{-1}(\mathcal{G}_{rs})$. Then $\pi_{rs} = \pi|_{\mathcal{G}_{rs}} \colon \widetilde{\mathcal{G}}_{rs} \to \mathcal{G}_{rs}$ is an unramified covering with Galois group W. In fact $\pi_{rs}^{-1}(X) = \{gwB \mid w \in W\}$, where gB satisfies $X \in adg\,\mathrm{Lie}\,B$.

(5.4) EXAMPLE. Take $\mathcal{G} = \mathfrak{gl}_2 = \left\{ \left(\begin{smallmatrix} a & b \\ c & d \end{smallmatrix} \right) \mid a, b, c, d \in \mathbf{C} \right\}$. Then $G = GL_2(\mathbf{C})$, B is the upper triangular group in G and G/B may be identified with the set of lines in $\mathbf{C}^2$, i.e., with $\mathbf{P}^1(\mathbf{C})$. Explicitly, the point in $\mathbf{P}^1(\mathbf{C})$ with homogeneous coordinates $\left[\begin{smallmatrix} x \\ y \end{smallmatrix} \right]$ is identified with gB where $g = \left(\begin{smallmatrix} x/y & 1 \\ 1 & 0 \end{smallmatrix} \right)$ if $y \neq 0$ or $g = I$ if $y = 0$. A short calculation shows that here

$$\widetilde{\mathcal{G}} = \left\{ \left(\left(\begin{smallmatrix} a & b \\ c & d \end{smallmatrix} \right), \left[\begin{smallmatrix} x \\ y \end{smallmatrix} \right] \right) \in A^4 \times \mathbf{P}^1 \mid cx^2 + (d-a)xy - by^2 = 0 \right\}.$$

Moreover $\mathcal{G}_{rs} = \left\{ \left(\begin{smallmatrix} a & b \\ c & d \end{smallmatrix} \right) \in \mathcal{G} \mid (d-a)^2 \neq -4bc \right\}$. Thus $\mathcal{G}_{rs}$ consists precisely of the points A in $\mathcal{G}$ such that $\pi^{-1}(A)$ consists of 2 points, where $\pi \colon \widetilde{\mathcal{G}} \to \mathcal{G}$ is the first projection.

To see how the Springer action on $H^*(\mathcal{B}_A)$ arises we work in the category $\mathcal{D}_c^b(Y)$ of bounded complexes of sheaves on Y, with constructible cohomology sheaves (see [BBD] or [Sp1]) for appropriate spaces Y (e.g., complex algebraic varieties).

If $\mathbf{C}$ is the constant sheaf on $\widetilde{\mathcal{G}}_{rs}$, then the direct image $(\pi_{rs})_* \mathbf{C} = \mathcal{L}$ is a local system (i.e., locally constant sheaf) on $\mathcal{G}_{rs}$ which admits the covering group W as a group of sheaf automorphisms. Now the intersection complex functor $IC^\bullet$ (see [BBD] or [Sp1]) may be applied to $\mathcal{L}$ to obtain a complex $IC^\bullet(\mathcal{G}, \mathcal{L}) \in \mathcal{D}_c^b(\mathcal{G})$.

Lusztig's key result is that

(5.5) $$IC^\bullet(\mathcal{G}, \mathcal{L}) \simeq \mathbf{R}\pi_* \mathbf{C} \quad \text{in } \mathcal{D}_c^b(\mathcal{G})$$

where $\mathbf{R}\pi_*$ is the direct image functor applied to the complex with $\mathbf{C}$ in degree 0 and 0 elsewhere.

The functor $IC^\bullet$ transfers the W-action from $\mathcal{L}$ to the left side of (5.5), and hence to $\mathbf{R}\pi_* \mathbf{C}$. It follows that W acts on the cohomology sheaves of $\mathbf{R}\pi_* \mathbf{C}$ and their stalks. Finally, observe that

$$\mathcal{H}^i(\mathbf{R}\pi_* \mathbf{C})_A = H^i(\pi^{-1}A, \mathbf{C}) = H^i(\mathcal{B}_A)$$

(since π is a proper map). This completes our sketch of the origin of the W-action on $H^i(\mathcal{B}_A)$.

In view of Theorem (5.3)(ii) the computation of the Green functions is largely reduced to computing the polynomials

(5.6)
$$Q_A = \sum_{i \geqslant 0} H^{2i}(\mathcal{B}_A)q^i \in R(W)[q]$$

where $R(W)$ is the Grothendieck ring of W, A is a nilpotent element of the complex semisimple Lie algebra $\mathcal{G}$ and $\mathcal{B}_A = \{ gB \in G/B \mid A \in \operatorname{ad}g \operatorname{Lie}B \}$.

The computation of the Q_A in turn reduces to the following.

(5.7) PROBLEM. *For each irreducible representation τ of W, compute the polynomial*

$$\langle Q_A, \tau \rangle = \sum_{i \geqslant 0} \langle H^i(\mathcal{B}_A), \tau \rangle q^i \in \mathbf{Z}[q].$$

Now the representations of W have a geometric description due to Springer [Sp2]. According to this description, each irreducible representation τ of W corresponds to a unique pair (A_τ, ϕ) where A_τ is a nilpotent element in $\mathcal{G}$ and ϕ is a character of $C(A_\tau)$ (the group of components of the centralizer of A_τ in G under the adjoint action).

(5.8) PROBLEM. *With the above notation, suppose the representation τ of W corresponds to (A_τ, ϕ) $(\phi \in C(A_\tau)\hat{\ })$. Find an explicit expression for $\langle Q_A, \tau \rangle$ which depends only on A, A_τ and ϕ.*

An alternative is

(5.9) PROBLEM. *Find an expression for $\langle Q_A, \tau \rangle$ which depends only on the geometry of the representation τ of W.*

There has been some progress in the solution of these problems and we outline currently known results and conjectures in the next section.

6. RESULTS CONCERNING THE POLYNOMIALS $\langle Q_A, \tau \rangle$

We begin this section by considering the case $A = 0$.

In this case $\mathcal{B}_A = \mathcal{B} = G/B$, the whole flag variety. If T is a maximal torus of G such that $T \subset B$, then we have a locally trivial fibration $G/T \to G/B$ with fibre U (where U is the unipotent radical of $B = TU$, and is a maximal connected unipotent subgroup of G). Hence $H_c^*(G/T) \cong H^*(G/B)$ up to a shift. But $W = N_G(T)/T$ acts on G/T (as $w.gT = gwT$) and hence on $H_c^*(G/T)$. Thus $H^*(G/B)$ inherits a W-action from the isomorphism above. This is called the 'classical' W-action on $H^*(G/B)$.

Now Borho and MacPherson have proved

(6.1.1) *The Springer action on $H^*(G/B)$ as defined above coincides with the classical action.*

For a proof, see [**Sh2**], §5.

The classical W-action on $H^*(\mathcal{B})$ is well understood [**Sp2**]. We have

(6.1.2) *Let S be the symmetric algebra on the Euclidean space on which W acts in its defining representation and let F be the ideal of S generated by invariants of positive degree. Then $H^{2i}(\mathcal{B}) \cong (S/F)_i$ is the ith graded component of S/F, and $H^{2i+1}(\mathcal{B}) = 0$, for all i.*

The proof of (6.1.2), which may be found in [**Sp2**], interestingly depends on the comparison theorem (3.1.3)(iii) above and an analysis of the action of $F \circ \mathrm{ad} w$ on $H^*_c(G(\overline{\mathbf{F}}_q)/T(\overline{\mathbf{F}}_q), \overline{\mathbf{Q}}_\ell)$, i.e., the proof depends on an argument which involves characteristic p. More precisely, one has

$$(6.1.3) \qquad \sum_i \mathrm{tr}\,(w, H^{2i}_c(G/T)\, q^{i+N} = |G^F/T_w^F|$$

where T_w is an F-stable torus of G which corresponds to $w \in W$ (cf. Lemma 4.3 above), and N is the number of positive roots. (Here we abuse notation by writing G for $G(\overline{\mathbf{F}}_q)$ etc. on the right hand side.)

This gives an explicit expression for the left hand side of (6.1.3), and comparison with a similar expression for $\sum_{i \geqslant 0} \mathrm{tr}(w, (S/F)_i) q^i$ yields (6.1.2). Thus the proof amounts to several applications of the Lefschetz principle.

In summary, we have, for $A = 0$,

$$(6.1.4) \qquad Q_0 = \sum_{i=0}^{N} (S/F)_i q^i$$

where $(S/F)_i$ is the W-module defined in (6.1.2).

(6.1.5) The inner products $\langle Q_0, \tau \rangle = f_\tau$ are called the *fake degrees*, and are known explicitly case by case.

In some cases there is a general closed formula; e.g., if ρ is the reflection representation of W, then

$$(6.1.6) \qquad \langle Q_0, \rho \rangle = \sum_{i=1}^{\ell} q^{m_i}$$

where $m_1, m_2, \ldots, m_\ell$ are the exponents of W. More generally, if ρ_j is the jth power of ρ (the 'jth compound of ρ') then

$$(6.1.7) \qquad \langle Q_0, \rho_j \rangle = \sigma_j(q^{m_1}, q^{m_2}, \ldots, q^{m_\ell})$$

where σ_j denotes the jth elementary symmetric function. The results (6.1.6) and (6.1.7) are due to Solomon [**Sol**].

There is a general analogue of (6.1.6) (see [**LS**] for a more precise statement):

(6.2) THEOREM (Lehrer-Shoji, Solomon, Spaltenstein). *Let ρ be the reflection representation and suppose the nilpotent element $A \in \mathcal{G}$ is of parabolic type. Then we have $\langle Q_A, \rho \rangle = \sum_{j \in S(A)} q^j$, where $S(A)$ is a set of integers associated with A, which arises in the geometry of the hyperplane complement associated with W.*

The next two general results are instrumental in the proof of (6.2). Suppose $P \supseteq B$ is a parabolic subgroup of G and write $\mathcal{P} = G/P$. Let $W(P)$ be the (parabolic) subgroup of W corresponding to P (i.e., $W(P)$ is the Weyl group of a Levi subgroup of P). Denote by $\mathcal{P}_A$ the analogue for P of $\mathcal{B}_A$, i.e., $\mathcal{P}_A = \{ gP \in \mathcal{P} \mid A \in \operatorname{ad} g \operatorname{Lie} P \}$.

(6.3) THEOREM (Borho-Macpherson [BM]). *With the above notation, we have*

$$\langle Q_A, \operatorname{Ind}_{W(P)}^W(1) \rangle = P(\mathcal{P}_A) = \sum_{i \geqslant 0} \dim H^{2i}(\mathcal{P}_A) q^i.$$

(6.4) THEOREM (Lehrer-Shoji [LS]). *Maintaining the above notation, let τ be an irreducible constituent of $\operatorname{Ind}_{W(P)}^W(1)$. Then*

$$\langle Q_0, \tau \rangle - P_e(\mathcal{P} \setminus \mathcal{P}_A) \leqslant \langle Q_A, \tau \rangle \leqslant \langle Q_0, \tau \rangle + P_o(\mathcal{P} \setminus \mathcal{P}_A)$$

where $P_e(Y) = \sum_{i \geqslant 0} \dim H_c^{2i}(Y) q^i$ and $P_o(Y) = \sum_{i \geqslant 0} \dim H_c^{2i+1}(Y) q^i$ for a complex algebraic variety Y.

In the statement (6.4) inequality for polynomials is defined coefficient-wise, i.e., $\sum a_i q^i \leqslant \sum b_i q^i$ if $a_i \leqslant b_i$ for each i. The inequality given relates $\langle Q_A, \tau \rangle$ to the fake degree $\langle Q_0, \tau \rangle$ and in practice gives much information concerning $\langle Q_A, \tau \rangle$.

(6.5) THEOREM (Borho-MacPherson) *Suppose $\tau \in \widehat{W}$ corresponds to the pair (A_τ, ϕ) (where A_τ is nilpotent in $\mathcal{G}$ and $\phi \in C(A_\tau)\widehat{}$ — see (5.8)). Then*
(i) $\langle Q_A, \tau \rangle = 0$ *unless $A \in \overline{((A_\tau))}$ where $((A_\tau))$ is the $\operatorname{ad} G$-orbit of A_τ and $\overline{((A_\tau))}$ is its closure in $\mathcal{G}$;*
(ii) $\langle Q_{A_\tau}, \tau \rangle = q^{d_A}$, *where $d_A = \dim Z_G(A)$.*

(6.6) THEOREM (Alvis-Lusztig). *Suppose A is regular nilpotent in a Levi subalgebra $\mathcal{L}$ of $\mathcal{G}$. Then*

$$(Q_A)_{q \to 1} = \operatorname{Ind}_{W(L)}^W(1),$$

where $W(L)$ is the parabolic subgroup of W which corresponds to $\mathcal{L}$.

We may now return to the empirical results of example (5.1) above and explain them in the light of the above results.

Clearly (5.1.1) is a special case of Theorem (5.3)(i). The leading coefficients of $Q_1(u)$ are, in the light of (5.3), the dimensions of the modules $H^{\text{top}}(\mathcal{B}_u)$, which in the case of GL_n are the irreducible W-modules. In general, we have

(6.7.1) *For A subregular, $Q_A = 1 + q\rho$ where ρ is the reflection representation of W.*

(5.1.3) is a special case of this.

The observation (5.1.4) concerning the sign character of W is a special case of (6.1.4), which asserts that $Q_0 = \sum_{i=0}^{N} (S/F)_i q^i$, because $(S/F)_N$ is known to be 1-dimensional, with basis the product of all the positive coroots, i.e., the unique alternating polynomial for W.

Finally the observation (5.1.5) is an instance of (6.5)(i), since the closure of the subregular class contains every nilpotent class except the regular one.

7. CONCLUDING REMARKS AND PROBLEMS

In § 3 several instances of the interplay of geometry, topology and group representations were pointed out, while in §§ 4, 5, 6 the relationship between the representation theory of the finite groups $G(\mathbf{F}_q)$ and the topology of the complex Lie groups $G(\mathbf{C})$ and their associated homogeneous spaces was discussed. Associated with many of the issues raised are interesting areas for further investigation. In this section, we point out several of these.

7.1. THE SPLIT BUILDING

Let G be a finite group of Lie type. Define the *split building* S of G to be the simplicial object associated with the poset whose elements are ordered pairs (P, Q) of opposite parabolic subgroups (the order relation being inclusion of the first term). This space is a covering of the Tits building T (of (3.2) above) and is known ([Ch]) to be spherical in type A.

(7.1.1) PROBLEM. *Study the geometry of the split building and the associated homology representations of G.*

This problem is under investigation by L. Rylands, and there is strong evidence that the split building is spherical in all classical types. The representation of $G(\mathbf{F}_q)$ on the top homology also has interesting properties.

7.2. PROBLEM

For any finite group G, study the relationship between the topology of its Brown (or Quillen) complex and the structure and representation theory of G (cf. [Qu], [Al]).

7.3. Hyperplane complements

Several instances of the relationship between the geometry of hyperplane complements and representations of $G(\mathbf{F}_q)$ have been mentioned above. A direct connection between these theories arises from the observation that the set of regular elements in a Cartan subalgebra of a semi-simple Lie algebra is precisely the complement of the hyperplanes corresponding to the coroots.

(7.3.1) PROBLEM. *Study the topology of* t_{reg}, *where* t *is a Cartan subalgebra of* $\mathcal{G}(\mathbf{C})$, *from the point of view of*
  (i) *Frobenius acting on* $t(\overline{\mathbf{F}}_q)$ *(cf. (3.1)), and*
  (ii) *intersection complexes on* t.
Relate this to the Springer representations.

7.4. Hecke algebras

Hecke algebras also play an important role in the representation theory of the groups $G(\mathbf{F}_q)$ (see [**Le3**]). Moreover it is clear that Hecke algebras have close connections with both the geometry of Schubert varieties (see [**DyL**]) and hyperplane complements (see [**Le3**]).

(7.4.1) PROBLEM. *Find a direct connection between the geometry of Schubert varieties and the geometry of hyperplane complements, from which Hecke algebras appear naturally.*

The work of Tanisaki [**T**] on the W-actions on scheme theoretic intersections of nilpotent orbits with a Cartan subalgebra may be thought of as a contribution to this problem. (I am indebted to P. Slodowy for this remark).

7.5. Green polynomials

We turn to the Green polynomials $Q_A \in R(W)[q]$ (see (5.6) above). The general problems have been stated above as (5.7), (5.8) and (5.9) and may be summarised as

(7.5.1) PROBLEM. *For each nilpotent element* $A \in \mathcal{G}(\mathbf{C})$ *and* $\tau \in \widehat{W}$, *find a geometric expression for* $\langle Q_A, \tau \rangle \in \mathbf{Z}_{\geqslant 0}[q]$.

In §6 several results concering these polynomials were given, which show that the computation of the polynomials $\langle Q_A, \tau \rangle$ is intimately connected with the cohomology theory of the homogeneous spaces G/P, where P is a parabolic subgroup in $G = G(\mathbf{C})$.

7.6. Problem

Investigate the implications of results such as (6.3) and (6.4) in both directions; i.e., what does a knowledge of the cohomology of G/P imply about the representations of $G(\mathsf{F}_q)$ and vice versa. In particular, for $G = GL_n$ every $\tau \in \widehat{W}$ is a linear combination of the representations $\mathrm{Ind}_{W(P)}^{W}(1)$. Deduce combinatorial results about flag varieties from (6.3).

7.7. Comment

The result (6.2) again points to a connection between hyperplane complements and the flag variety of G. This reinforces the question asked in (7.4.1).

7.8. Problem

Finally, here is a more specific question. The results (6.1.7) and (6.2) suggest the following.

(7.8.1) Conjecture. *Let ρ_j be the jth exterior power of the reflection representation $\rho = \rho_1$ of W. Then for a nilpotent element $A \in \mathcal{G}$ of parabolic type we have*

$$\langle Q_A, \rho_j \rangle = \sigma_j(q^k, q^\ell, q^m, \dots)$$

where $S(A) = \{k, \ell, m, \dots\}$ is the set of integers referred to in (6.2), and σ_j is the jth elementary symmetric function.

This has been verified for several low-dimensional cases, but at present is known in general only for $A = 0$ (see (6.1.7) above) and for $j = 1$ (see (6.2)).

Finally, we mention the important work of Lusztig (see [L2] and the references there) on character sheaves. This work also represents a 'geometrization' of the representation theory of the groups $G(\mathsf{F}_q)$; an orthonormal basis of the space of class functions is constructed, which is closely related to the set of characters. The functions appear as characteristic functions of perverse sheaves on $G(\overline{\mathsf{F}}_q)$. This work is closely related to the problems discussed above.

References

[A1]   J. L. Alperin, 'A Lie approach to finite groups', in these Proceedings.

[BBD] A. A. Beilinson, J. Bernstein and P. Deligne, 'Faisceaux pervers', *Astérisque* 100 (1982).

[BM]   W. Borho and R. MacPherson, 'Partial resolutions of nilpotent varieties', *Astérisque* 101–102 (1983), 23–74.

[B]   N. Bourbaki, *Groupes et algèbres de Lie, Ch. IV, V, VI*, (Hermann, Paris, 1968; Masson, Paris, 1981).

[Br1] E. Brieskorn, 'Sur les groupes de tresses [d'après V. I. Arnold]', in *Séminaire Bourbaki*, vol. 1971/2, Exp. No. 401; Lecture Notes in Math. 317, pp. 21–44 (Springer-Verlag, Berlin Heidelberg New York, 1973).

[Br2]   E. Brieskorn, 'Singular elements of semi-simple algebraic groups', in *Actes Congrès Intern. Math.* (*Nice,* 1970), vol. 2, pp. 279–284 (Gauthier-Villars, Paris, 1971).

[Ch]   Ruth Charney, 'Homology stabiltiy for GL_n of a Dedekind domain', *Invent. Math.* **56** (1980), 1–17.

[CLT]   C. W. Curtis, G. I. Lehrer and J. Tits, 'Spherical buildings and the character of the Steinberg representation', *Invent. Math.* **58** (1980), 201–210.

[CL1]   C. W. Curtis and G. I. Lehrer, 'Homology representations of finite groups of Lie type', *Contemp. Math.* **9** (1981), 1–28.

[CL2]   C. W. Curtis and G. I. Lehrer, 'A new proof of a theorem of Solomon-Tits', *Proc. Amer. Math. Soc.* **85** (1982), 154–156.

[D]   P. Deligne, *Cohomologie étale* (SGA $4\frac{1}{2}$), Lecture Notes in Math. **569** (Springer-Verlag, Berlin Heidelberg New York, 1977).

[DL]   P. Deligne and G. Lusztig, 'Representations of reductive groups over finite fields', *Ann. of Math.* (2) **103** (1976), 103–161.

[DyL]   M. J. Dyer and G. I. Lehrer, 'On positivity in Hecke algebras', *Geom. Dedicata* (1990) (to appear).

[JAG]   J. A. Green, 'The characters of the finite general linear groups', *Trans. Amer. Math. Soc.* **80** (1955), 402–447.

[H]   Graham Higman, 'Representations of general linear groups and varieties of p-groups', in *Proc. Internat. Conf. Theory of Groups,* Canberra, 1965; ed. by L. G. Kovács and B. H. Neumann, pp. 167–173 (Gordon and Breach, New York, 1967).

[Ka]   D. Kazhdan, 'Proof of Springer's hypothesis', *Israel J. Math.* **28** (1977), 272–286.

[Le1]   G. I. Lehrer, 'On the Poincaré series associated with Coxeter group actions on complements of hyperplanes', *J. Lond. Math. Soc.* (2) **36** (1987), 275–294.

[Le2]   G. I. Lehrer, 'On hyperoctahedral hyperplane complements', in *The Arcata conference on representations of finite groups,* ed. by Paul Fong, Proc. Symposia in Pure Math. vol. 47 Part 2, pp. 219–234 (Amer. Math. Soc., Providence, 1987).

[Le3]   G. I. Lehrer, 'The spherical building and regular semisimple elements', *Bull. Austral. Math. Soc.* **27** (1983), 361–379.

[Le4]   G. I. Lehrer, 'A survey of Hecke algebras and the Artin braid group', *Contemp. Math.* **78** (1988), 365–385.

[LS]   G. I. Lehrer and T. Shoji, 'On flag varieties, hyperplane complements and Springer representations of Weyl groups', *J. Aust. Math. Soc. Ser. A* (to appear).

[LSo]   G. I. Lehrer and Louis Solomon, 'On the action of the symmetric group on the cohomology of the complement of its reflecting hyperplanes', *J. Algebra* **104** (1986), 410–424.

[L1]   G. Lusztig, *The discrete series of GL_n over a finite field,* Annals of Math. Studies **81** (Princeton University Press, Pinceton, 1974).

[L2]   G. Lusztig, 'Green functions and character sheaves' (preprint, 1989).

[L3]   G. Lusztig, 'Green polynomials and singularities of nilpotent classes', *Adv. in Math.* **42** (1981), 169–178.

[Mac]   I. G. Macdonald, *Symmetric functions and Hall polynomials* (Clarendon Press, Oxford, 1979).

[OS]   P. Orlik and L. Solomon, 'Combinatorics and the topology of complements of hyperplanes', *Invent. Math.* **56** (1980), 167–189.

[Qu]   D. Quillen, 'Homotopy properties of the poset of nontrivial p-subgroups of a group', *Adv. in Math.* **28** (1978), 101–128.

[Sh1]   T. Shoji, 'On the Green functions of classical groups', *Invent. Math.* **74** (1983), 239–267.

[Sh2]   T. Shoji, 'Geometry of orbits and Springer correspondence', *Astérisque* **168** (1988), 61–140.

[Sol]   L. Solomon, 'Invariants of finite reflection groups', *Nagoya Math. J.* **22** (1963), 57–64.

[Sp1] T. A. Springer, 'Perverse sheaves and representation theory', in *The Arcata conference on representations of finite groups*, ed. by Paul Fong, Proc. Symposia in Pure Math. vol. 47 Part 1, pp. 315–322 (Amer. Math. Soc., Providence, 1987).

[Sp2] T. A. Springer, 'Trigonometric sums, Green functions of finite groups and representations of Weyl groups', *Invent. Math* **36** (1976), 173-207.

[SSt] T. A. Springer and R. Steinberg, 'Conjugacy classes', in *Seminar on algebraic groups and related finite groups*, ed. by A. Borel et al., Lecture Notes in Math. **131**, pp. 167–266 (Springer-Verlag, Berlin Heidelberg New York, 1970).

[St] R. Steinberg, *Endomorphisms of linear algebraic groups*, Mem. Amer. Math. Soc. **80**, 1968.

[T] Toshiyuki Tanisaki, 'Defining ideals of the closures of the conjugacy classes and representations of the Weyl groups', *Tôhoku Math. J.* (2) **34** (1982), 575–585.

[PW] P. Webb, 'Subgroup complexes', in *The Arcata conference on representations of finite groups*, ed. by Paul Fong, Proc. Symposia in Pure Math. vol. 47 Part 1, pp. 349–365 (Amer. Math. Soc., Providence, 1987).

[AW] A. Weil, 'Numbers of solutions of equations in finite fields', *Bull. Amer. Math. Soc.* **55** (1949), 497–508.

Department of Pure Mathematics
UNIVERSITY OF SYDNEY, NSW 2006
Australia

SOME INTERACTIONS BETWEEN GROUP THEORY AND THE GENERAL THEORY OF ALGEBRAS

RALPH MCKENZIE

I was invited to give a survey talk at the Neumann Conference on "areas of contact between my interests and group theory." I found the atmosphere of friendly collegiality that prevailed during the entire conference week very impressive. It was a very special conference—the best sort of tribute to Bernhard Neumann. The high esteem and affection felt for him by his students and colleagues were perfectly evident. The second thing to impress me about this conference was the quality of the lectures: it seemed that I had never attended a conference where the lectures, almost without exception, were so well prepared, and so full of substance.

While these impressions matured in my mind, I had three days to consider once again what I might say, drawn from my own twenty-five years of experience in mathematical research outside of group theory, that might be of interest to this gathering of group theorists. This paper is an account of some mathematical developments in which groups played a role, mostly selected from my own research experience, but including some developments that I admired from a distance. It is a much expanded and much altered version of the remarks I delivered to the Conference.

Although I began my research career specializing in logic—under the influence of Donald Monk at the University of Colorado, and later under Monk's teacher Alfred Tarski at Berkeley—all the research I've done and all the mathematics I'm competent to write about falls under the heading of "universal algebra and lattice theory". This field is essentially the same as what Tarski liked to call "the general theory of algebras". It seems to me that there are deep underlying affinities between the general theory of algebras and the theory of groups that offer promise of a substantial dialogue in the future, as the general theory of algebras continues to develop. While relating this general (or universal) algebra to group theory, I will attempt to convey an accurate picture of what the field consists of in its present state of development.

What is universal algebra? Universal-algebra-and-lattice-theory is a thoroughly modern subject, whose origins can be traced to two papers of Garrett Birkhoff, [1933] and [1935]. It is one of the youngest of mathematical branches, on a par with semigroup theory; and currently it is the scene of a flourishing mathematical enterprise.

Its literature contains more than 1500 papers published since 1970, and its community of scientific workers includes, worldwide, I estimate, around 300 members, most of whom reside in the United States, in central Europe, and in the Soviet Union. In earlier times, the subject received contributions from mathematicians of diverse stripes, including logicians like Alfred Tarski and Roger Lyndon (in his early years), and algebraists like Philip Hall, Graham Higman, A. G. Kurosh, A. I. Mal'cev (who, among other accomplishments, founded a flourishing Soviet school of general algebra), and Bernhard Neumann.

Contributions from outside the community are still occurring, though less frequently. The field has undergone significant growth and transformation during the past two decades, and demands proportionately greater commitment and specialization from those who will leave their mark on it. I cannot expect that many of my readers will have had any occasion to become familiar with the current aims, methods, and research directions in universal algebra; therefore, my aim in the next few paragraphs will be to give some idea of the mathematical universe as universal algebraists see it, while sharply distinguishing our outlook from the perspectives of the closely related fields of category theory and logic (or model theory).

Universal algebraists tend to vehemently reject any suggestion that our subject is properly a part of logic, or of category theory (which places severe limits on the possibilities for communication with certain logicians and category theorists). But in our viewpoints on mathematics, we do have much in common with these other fields; and the three fields share much common ground. In my view, our subject is properly classified as an independent branch of algebra. It is an adolescent branch, but, as I hope to show, one that has already developed far beyond the stage where it could be characterized (as it once was) as consisting of a collection of the most basic and general results that hold simultaneously for diverse systems such as groups, rings, and modules.

Algebras. The universal algebraist has a peculiar vision and point of view that set him apart from specialists in logic and category theory or any other field of mathematics. He considers that every mathematical system in which the structure is defined by a set together with a collection of finitary operations defined over the set, is potentially interesting and deserving of study. If the tools of logic, category theory, group theory or any other branch offer promise for his investigations of algebras, he will use them. But he is not committed to any traditional approach. In fact, he hopes to find completely new kinds of algebras; interesting types of structure in these algebras that is absent or not clearly revealed in groups, rings, and modules; and completely new techniques for revealing the structure. A universal algebraist differs from a category theorist in having no commitment to translate every idea into the language of categories, maps, and functors; from the logician in having no commitment to the types of questions that are amenable to the techniques of ultraproduct constructions and logical compactness arguments. Indeed, he is likely to be especially interested in finite algebras, and passionately committed to the investigation of certain types of questions concerning his "algebras" that, to a category theorist, a logician, or a classically trained algebraist, would appear to be either obscure, misguided, uninteresting, or just hopelessly intractable. You should have an idea what I mean after reading this article through to its conclusion.

There is one striking point of similarity in the attitudes of logicians, universal algebraists and category theorists. They share a modern concern to achieve the utmost generality. An incredible diversity of complex systems is implicitly admitted for consideration and the only actual limitation on the generality of results obtained is the natural desire to achieve results of some depth. This emphasis on examining an almost unlimited spectrum of diverse complex systems is, perhaps unintentionally, in conformance with an obvious quality of the world we live in; and it is, I think, a virtue which will help these disciplines to survive into the twenty-first century and beyond.

Lattices. In the general theory of algebras, lattice theory emerges, more clearly than in other branches of algebra, as an important tool for the structural analysis of algebras. The congruence lattice is a very important invariant of general algebras. Its study has often led to the deepest and most interesting results. Thus universal algebra has supplied a major impetus for the development of lattice theory; and the two disciplines are very closely linked. Lattices are important in universal algebra for another reason. They constitute the most important class of algebras exhibiting interesting algebraic phenomena radically different from what one encounters in groups, rings, or modules.

Varieties. Another constant in almost all current work in universal algebra is the central place accorded to varieties. In this tradition, *variety* means the same thing as in H. Neumann's book *Varieties of Groups*. It is a class of algebras, all with the same type of operations, that is closed under the formation of direct products, subalgebras, and homomorphic images. This concept of a variety was defined in one of the first contributions to universal algebra, G. Birkhoff [1935], where several fundamental results about varieties were proved, including the existence of free algebras in any variety and the fact that varieties are the same thing as classes of algebras defined by sets of equations.

Mal'cev conditions. There are numerous theorems that assert that all algebras in a variety possess a certain property if and only if there are terms in the language of the variety and certain equations between these terms that are valid equations in all the algebras of the variety. The paradigm for all these theorems is the 1954 result of A. I. Mal'cev according to which the algebras in a variety possess permuting congruences if and only if there is a term $t(x,y,z)$ such that the equations $t(x,y,y) \approx x$ and

$t(y, y, x) \approx x$ are valid in the variety. The variety of groups has such a term—take, for example, $t(x, y, z) = x \cdot y^{-1} \cdot z$. Any condition of this sort, asserting the existence of terms obeying some equations, is called a *Mal'cev condition* and the family of varieties satisfying it is called a *Mal'cev class*. The classification of varieties by dividing them into Mal'cev classes was a dominant theme of research in universal algebra during the entire decade of the 1970's.

I believe that nobody is very happy with the name "universal algebra" which seems to have stuck to the subject. To avoid publishing yet another book with this title, I and my co-authors G. McNulty and W. Taylor chose to incorporate the three principal themes of the subject into the title of our recent book: "Algebras, Lattices, Varieties." You will notice that lattices and varieties are visibly present in most of the problems and results related in this article.

Commutator theory and tame congruence theory. During the past decade, two theories created by universal algebraists have attained a level of power and generality not seen before in this field. The first, *general commutator theory*, applies to all algebras **A** that share with groups the property that the congruence lattices of **A** and of all algebras in the variety generated by **A** are modular lattices; i.e., it applies to algebras that belong to congruence-modular varieties. So it applies to groups, rings, and modules, but also to lattices, quasigroups, Heyting algebras and many less-familiar algebras.

The commutator is simply a binary operation on the congruence lattice. It is a generalization of the operation of forming the commutator subgroup $[N, K]$ from two normal subgroups N and K of a group **G**; and it coincides with this operation in groups (after identifying congruences with normal subgroups). Moreover, it possesses all the abstract properties that can be proved to hold for the commutator of normal subgroups. We write $[\theta, \psi]$ for the commutator of two congruences θ and ψ. Then for any set $\{\theta, \psi, \delta\} \cup \{\psi_i : i \in I\}$ of congruences on an algebra **A** that belongs to a congruence-modular variety, the commutator is symmetric and sub-multiplicative, i.e., $[\theta, \psi] = [\psi, \theta] \subseteq \theta \cap \psi$; it is join-distributive, i.e., $[\theta, \bigvee\{\psi_i : i \in I\}] = \bigvee\{[\theta, \psi_i] : i \in I\}$; and the commutator of congruences of a quotient algebra $\mathbf{A}/\delta$ is determined from commutators in **A** by the rule $[\theta/\delta, \psi/\delta] = ([\theta, \psi] \vee \delta)/\delta$, if $\delta \subseteq \theta \cap \psi$. Restricted to the algebras in a fixed congruence-modular variety, this commutator is the "largest" binary congruence operation that possesses all of these properties.

The concepts of an Abelian algebra **A** (i.e., one satisfying $[1_A, 1_A] = 0_A$ where 1_A and 0_A are the largest and least congruences of **A**) and of an Abelian congruence β (satisfying $[\beta, \beta] = 0_A$), and the notions of solvable and nilpotent congruences and algebras, are defined in this theory; and the behaviour of these concepts in groups furnishes a reliable guide to their behaviour in this very general context. In particular, an algebra is Abelian if and only if it is polynomially equivalent with a module over a ring. General commutator theory and many of its applications are developed in the book R. Freese, R. McKenzie [1987]. Applications have ranged from unique factorization results for algebras in congruence-modular varieties, and structural characterizations of locally finite congruence-modular varieties with decidable first-order theory, to a proof that residually small congruence-modular varieties with the amalgamation property have enough injectives, and a better understanding of residually small varieties (a topic to which we shall return below).

General commutator theory is among the best achievements of general algebra to date; and with hindsight it can be seen to be implicit already in the elementary part of group theory that belongs to the common culture of all algebraists. The lattice of normal subgroups of a group, with the commutator operation, is a lattice ordered monoid. The concept of a lattice ordered monoid, which arose naturally in ideal theory, had been studied by W. Krull, G. Birkhoff, R. P. Dilworth and M. Ward, and by many others. But in these ongoing axiomatic studies there was no hint that a commutator could be naturally defined in such a broad context as congruence-modular varieties, and that it would have such power in applications. The general commutator theory was first developed (in a slightly less general form that applied only to algebras in varieties with permuting congruences) by J. D. H. Smith, who at the time was a graduate student working with J. Conway. (See J. D. H. Smith [1976].) The full development for modular varieties was achieved by J. Hagemann and C. Herrmann [1979].

Residually small varieties. Suppose that $\mathbf{G}$ is a group and that the variety generated by $\mathbf{G}$ is residually small, i.e., there is a cardinal number λ such that every group $\mathbf{K}$ in this variety can be embedded into a product of groups of cardinality $< \lambda$. What does this tell us about $\mathbf{G}$? This seems to be a difficult question. A. Yu. Ol'shanskii [1960] proved that if $\mathbf{G}$ is finite, then the variety it generates will be residually small just in case all the nilpotent groups in this variety are Abelian; and this holds if and only if the Sylow subgroups of $\mathbf{G}$ all are Abelian. Ol'shanskii also proved that every locally finite residually small variety of groups is generated by some one finite group. In one of the first substantial applications of general commutator theory, R. Freese and R. McKenzie [1981] proved that if $\boldsymbol{V}$ is any residually small congruence-modular variety of algebras, then $\boldsymbol{V} \models x \wedge [y, y] = [x \wedge y, y]$, where (x, y) is understood to be a pair of variables ranging over pairs of congruences of any algebra in $\boldsymbol{V}$. They also showed that if $\mathbf{A}$ is a finite algebra in a congruence-modular variety, then $\boldsymbol{HSP}(\mathbf{A})$ is residually small if and only if the subalgebras of $\mathbf{A}$ satisfy this commutator equation. R. McKenzie [1982b] proved that this commutator equation is equivalent to residual smallness for varieties of rings.

For non-locally-finite varieties of groups, to the best of my knowledge, not much is known about the consequences of residual smallness, other than the fact that the commutator equation of the last paragraph must hold; and moreover, a residually small variety of groups is residually $\leq 2^{\aleph_0}$ (which follows from a completely general result of W. Taylor [1972]). Recently, M. V. Sapir and L. N. Shevrin [1988] obtained a characterization of all residually small varieties of semigroups, modulo the characterization of residually small varieties of groups of finite exponent, thus completing investigations that had been begun by E. A. Golubov and M. V. Sapir [1979] and by R. McKenzie [1981] and [1983]. Perhaps some group theorist would find a worthy challenge in the determination of all residually small varieties of groups.

The RS-conjecture. One of the most difficult problems in universal algebra takes the form of a conjecture, known as the RS-conjecture. It is the statement that if a finite algebra **F** generates a residually small variety, then that variety must be residually bounded by some finite cardinal. This has been proved to be the case for algebras in congruence-modular varieties (using the commutator), for semigroups, and more recently for algebras in varieties satisfying any Mal'cev condition that does not hold in the variety of semilattices. Thus a counterexample, if it exists, must be a ridiculously unorthodox algebra.

The other important new theoretical development in the realm of universal algebra is *tame congruence theory*. This theory seems a more radical departure than commutator theory and I know of no models or precedents for it in mainstream algebra. Finite group theorists should be comfortable with this topic, as the dominant motifs are "finite algebras", "structure" (of course), and "classification". Unfortunately, this theory seems to be at too general a level to yield any new information about finite groups, although I will discuss that possibility a bit later. The theory was developed by myself and my former graduate student David Hobby, largely in the hope of using it to resolve the RS-conjecture. Our work has been published in D. Hobby and R. McKenzie [1988]. I call the theory "Tame Congruence Theory" (write TCT) because congruences of general algebras have always seemed to me to be rather obdurate and difficult objects to deal with, and seen through the lens of this theory, they become almost tame. TCT deals strictly with finite algebras (though it has some very nice applications to infinite locally finite algebras), but applies unrestrictedly to all finite algebras. TCT shows that every finite algebra can be usefully regarded as a sort of amalgam of algebras belonging to five basic classes. The five classes are: (1) G-sets, or finite algebras consisting of a finite set A and a collection of unary operations constituting a primitive group G of permutations on A; (2) finite vector spaces (over finite fields); (3) the two-element Boolean algebra; (4) the two-element lattice; (5) the two-element semi-lattice.

Let **A** be a finite algebra and $\langle \alpha, \beta \rangle$ be a pair of congruences of **A**, and assume that $\alpha \prec \beta$, i.e., that α is a proper subset of β and there is no congruence lying properly between the two. Such a pair $\langle \alpha, \beta \rangle$ is the same thing as a *prime quotient* in the lattice **Con A** of congruences of **A**. For example, **A** might be a simple algebra, in which case α and β are its only congruences. We associate with the pair a set of subsets of A, denoted $M_A(\alpha, \beta)$. The members of $M_A(\alpha, \beta)$, called $\langle \alpha, \beta \rangle$-*minimal sets*, are the minimal members in the collection of subsets $M \subseteq A$ such that for some polynomial function f of the algebra A, $f(A) = M$, and there exists at least one pair of β-congruent elements x and y such that $(f(x), f(y)) \notin \alpha$. In TCT, we show that each set $M \in M_A(\alpha, \beta)$ is of the form $e(A)$ for some idempotent polynomial function e (i.e., satisfying $ee = e$) and that for every $K \in M_A(\alpha, \beta)$ there are polynomial functions that induce a bijection from M onto K, and its inverse map from K onto M. We define the induced algebra, $A|_M$, to be the set M together with all operations on M that are restrictions of polynomials of A (of an arbitrary number of variables) under which M is closed. All of the induced $\langle \alpha, \beta \rangle$-minimal algebras $A|_M$ are then isomorphic to one another (actually, crypto-morphic) via polynomial bijections. We define the $\langle \alpha, \beta \rangle$-*traces* to be the sets N which, for some $M \in M_A(\alpha, \beta)$, are identical with some $\beta|_M$-equivalence class that contains at least two $\alpha|_M$-equivalence classes. The algebras induced on the $\langle \alpha, \beta \rangle$-traces are all isomorphic to one another via polynomial maps, and up to crypto-morphism, we have a unique algebra $(A|_N)/\alpha|_N$ associated with $\langle \alpha, \beta \rangle$ which belongs to one of the five classes of basic algebras defined in the last paragraph. In this way, the prime quotients in **Con A** get classified into five types; and the Hasse diagram of the finite lattice **Con A** becomes a labelled graph, each prime quotient labelled with one of the five integers $1, \ldots, 5$.

I was fascinated to learn from C. Praeger's talk at this conference that the prime quotients of type **1**, correlated with primitive permutation groups, can be further subdivided into six types.

The collections of $\langle \alpha, \beta \rangle$-minimal sets and $\langle \alpha, \beta \rangle$-traces have some further nice properties I have not mentioned. (It is useful to think of the traces as constituting a reasonably well-behaved "pseudo-geometric" structure on the universe of the algebra.) Using these properties, we find that the labelled congruence lattice of a finite algebra conveys considerable information about the structure of the algebra (while the labelled congruence lattices of the subalgebras of finite direct powers of the algebra give a good deal more information).

TCT has already found applications to many diverse topics, including: the classification of locally finite varieties by the Mal'cev-type conditions they satisfy and especially the classification by properties of the congruence lattices (there is a largest proper family of locally finite varieties definable by a Mal'cev condition with idempotent operations, and it consists of the varieties in which any pair of locally solvable congruences permute); the spectra of cardinalities of free algebras in locally finite varieties; residually small varieties (any residually small locally finite variety satisfying a Mal'cev condition involving idempotent operations that is not satisfied by the variety of semilattices, is congruence-modular); the spectra of simple algebras in locally finite varieties; forbidden lattices of varieties; and the structure of decidable varieties.

Decidable varieties. Probably the most substantial application of tame congruence theory to date is in the monograph by R. McKenzie and M. Valeriote [1989], where it is the chief tool used in showing that any decidable locally finite variety decomposes as the varietal product of three very special varieties, S having uniform type 1, A having uniform type 2, and D having uniform type 3. S is equivalent to a class of categories (in the usual meaning of the word) having a fixed finite number of objects. A is equivalent to the variety of all modules over a certain finite ring. D is a discriminator variety, which means that it is very like the variety of Boolean algebras in many respects.

This result contains an algorithm for reducing the question "is the variety generated by the finite algebra A decidable?" to the question "is the variety of modules over the finite ring R decidable?" and incidentally, the characterization of the finite rings with this property is still unknown, and it is not known whether this class of finite rings is a recursive class.

I suppose that it is relevant to mention that my interest in the structure of decidable varieties was initially stimulated by investigations into decidable varieties of groups carried out by Yu. L. Ershov and A. P. Zamjatin during the 1970's. Zamjatin proved the decisive result: a variety of groups is decidable if and only if it consists of Abelian groups.

Finite groups and finite lattices. Although tame congruence theory seems to have no obvious antecedents, in fact it grew out of some ideas contained in a paper on finite groups and finite lattices. In P. P. Pálfy and P. Pudlák [1980], the authors proved that every finite lattice is isomorphic to the congruence lattice of some finite algebra if and only if every finite lattice is isomorphic to an interval in the lattice of subgroups of some finite group.

The method of the Pálfy-Pudlák proof involved showing that every finite lattice is embedded as an interval in a finite lattice with some special properties, and if A is a finite algebra whose congruence lattice L has these special properties, then the induced algebra on some "minimal set" in A still has L as its congruence lattice (isomorphically) and the polynomial maps of the minimal algebra, minus the constant maps, form a transitive group of permutations. Tame congruence theory involves a refinement of the notion of minimal set introduced in that paper, together with a considerable elaboration of the properties of these sets.

In his talk at the Neumann Conference, Pálfy indicated his belief that, in fact, it is not true that every finite lattice is isomorphic to an interval in the subgroup lattice of a finite group. Some very simple lattices have proved to be difficult test cases for this problem; viz., the lattices of height two. M_n has $n + 2$ elements including its least and largest elements, 0 and 1, and n atoms which pairwise meet to 0 and join to 1. If $n - 1 = q$ is a prime power, then M_n is (isomorphic to) the lattice of subspaces of a 2-dimensional vector space over a q-element field; and so it is isomorphic to an interval in the subgroup lattice of the group generated by translations and rotations of vectors. The least value of n for which $\mathbf{M}_n$ admits no representation of this kind is $n = 7$.

Only very recently has it been possible, due to the classification theorem on finite simple groups and the resulting improvement in our knowledge of primitive permutation groups of odd degree, to show that $\mathbf{M}_7$, and also $\mathbf{M}_{11}$, are isomorphic to intervals at the top of the subgroup lattice of the alternating group of degree 31. This was done by W. Feit and P. P. Pálfy. Also, P. P. Pálfy claims to have verified that no larger $\mathbf{M}_n$ can be represented in this way, as an interval in the subgroup lattice of an alternating group of odd degree. So we have this special case of the problem still open: Does there exist a finite group $\mathbf{G}$ and a subgroup H for which there are precisely thirteen proper subgroups that properly contain H, all of them maximal subgroups of which any pair intersect to H?

Incidentally, among the surprising initial results to come out of tame congruence theory was the fact that $\mathbf{M}_7$ is not isomorphic to the congruence lattice of any finite algebra with just one operation, and not isomorphic to the lattice of subvarieties of any locally finite variety. Subsequently, W. Lampe [1986] improved the latter result by removing the adjective "locally finite."

Reducing very general problems to groups and rings. We have just seen examples of general questions about all finite algebras which universal algebra has reduced to questions about finite rings and finite groups, and which look to be very difficult to resolve in these classical domains. I mean the characterization of decidable locally finite varieties and the question of representing finite lattices as congruence lattices of finite algebras. Another such example can be found in R. McKenzie [1982a], where the characterization of finite algebras that generate a variety containing only a finite number of directly indecomposable algebras is reduced to the problem of characterizing the finite rings of finite representation type.

Congruence equations of finite groups. It may be that tame congruence theory will never have a measurable impact on finite group theory. Nevertheless, in this connection there are some further questions I would like to mention. Congruences of a group correspond to normal subgroups, so we deal with the labelled structure lattice $\mathbf{L}(\mathbf{G})$, i.e., the lattice of normal subgroups of the finite group $\mathbf{G}$ where the prime quotients have labels supplied by the theory. Since groups have permuting congruences (due to the fact that $K \cdot L = L \cdot K$ holds for $K, L \in \mathbf{L}(\mathbf{G})$), the possibilities are restricted. For a prime quotient $\langle K, L \rangle$, the type is $\mathbf{2}$ if $[L, L] \subseteq K$ and is $\mathbf{3}$ if L is not Abelian over K. In the second case, the minimal sets are some of the two-element sets $\{c, d\} \subseteq G$ for which $c \neq d$; this case seems totally uninteresting.

But the Abelian prime quotients are more interesting. When L is Abelian over K, the minimal sets $M \in \mathrm{M}_{\mathbf{G}}(K, L)$ have a prime-power cardinality and the induced algebras $G|_M$ are polynomially equivalent to loops with operators that are nilpotent in the sense of the general commutator theory. So these type-two $\langle K, L \rangle$-minimal algebras in groups are very special indeed; moreover, they look to be crucial for the study of $L(G)$. If we map congruences of G to the system of their restrictions to minimal sets, choosing one minimal set for each prime quotient of congruences, we obtain an isomorphism of $L(G)$ with a subdirect product of two-element lattices and congruence lattices of the type-two induced algebras. The snag is that we have no method for finding the $\langle K, L \rangle$-minimal sets and determining the congruence lattice of the induced loop with operators, other than by first computing the group of all unary polynomial functions of G (which is the subgroup of G^G generated by the constant functions and the identity function).

But why would anyone be interested in this sort of approach to examining the structure lattice of a finite group, by breaking it down as a subdirect product of congruence lattices of special but rather mysterious algebras? Well, if you are a group theorist, perhaps you will never be interested in this. In universal algebra, we have had several decades of experience attempting to compare and classify all kinds of algebraic systems by distinguishing special properties of the varieties they generate. Our experience seems to have taught us that at the most general level, the significant properties of varieties generally involve conditions on the congruence lattices of their algebras, such as congruence modularity, congruence distributivity, or congruence semi-distributivity.

So we formulated the concept of the congruence variety of a variety V, or CON V, which is the variety of lattices generated by the class of congruence lattices of algebras in V. We also have the congruence quasivariety, Q-CON V, associated to V, which is the class of lattices that can be embedded into the congruence lattice of some algebra in V. And we found that it is much easier to ask obvious questions about these congruence varieties and congruence quasivarieties than it is to answer them. For example, where G is the variety of groups, CON G and Q-CON G remain very mysterious.

For the class A of all Abelian groups, recursive sets of first-order axioms for Q-CON A and CON A have been given. B. Jónsson asked, more than thirty years ago, whether CON G = CON A, and this question is still open.

We don't know whether there is any finite group G for which $L(G)$ lies outside of CON A; and I observed several years ago that the above-outlined approach to $L(G)$ via tame congruence theory seemed to offer some promise for proving that in fact there is no such finite group. Then I discovered that by taking the restrictions of normal subgroups of G into the Sylow subgroups of G, we get a lattice embedding of $L(G)$ into a product of the lattices $L(S)$, S ranging over the Sylow subgroups. [I learned at this conference that P. Neumann had noticed this fact many years ago. This embedding is not as well-structured as the one occurring in TCT, in that we get a sublattice of the product lattice that may very well fail to project onto each factor. The failure is most obvious if $\mathbf{G}$ is a simple group.]

Now if $\mathbf{S}$ is a finite p-group, for example a Sylow subgroup of $\mathbf{G}$, then the minimal sets corresponding to prime congruence quotients of $\mathbf{S}$ are all equal to S. So for finite p-groups, tame congruence theory is completely trivial. I had to admit that these observations seem to imply that, as far as the Jónsson question $\mathrm{CON}\,\mathbf{G} = \mathrm{CON}\,\mathbf{A}$ is concerned, TCT offers no advantages.

But meanwhile, E. Kiss in Budapest had taken up my suggestion to consider the Jónsson question in connection with tame congruence theory, and he was considering some very small 2-groups as tractable examples of the so-called E-minimal algebras of type two associated with Abelian prime quotients, whose congruence lattices he wished to analyze. He could show that the eight-element groups are Abelianizable: each of them supports a binary operation (that is more or less unrelated to the operations of the given group) which makes the universe into an Abelian group whose congruences include the congruences of the given group. Perhaps he hoped to show that every finite p-group is Abelianizable. (I believe that group theorists have long known this to be false.) But he got stuck on some larger groups and turned to P. P. Pálfy for assistance. Pálfy had never considered the Jónsson question before, but now he weighed in with a very nice result: If $\mathbf{P}$ is the free group on three generators in the variety generated by the eight-element non-Abelian groups, then $\mathbf{L}(\mathbf{P})$ does not belong to Q-CON$\mathbf{A}$!

I was visiting Budapest and Darmstadt in summer 1988 and had the opportunity to convey Pálfy's handwritten notes containing his proof to C. Herrmann, an outstanding specialist in the study of modular lattices. Within several days, Herrmann was looking at a finite 2-group $\mathbf{Q}$ closely related to Pálfy's group; and he quickly convinced himself that he could prove both that $\mathbf{L}(\mathbf{Q}) \notin$ Q-CON$\mathbf{A}$ and that $\mathbf{L}(\mathbf{Q})$ is a projective modular lattice, from which it would follow that $\mathbf{L}(\mathbf{Q}) \notin \mathrm{CON}\,\mathbf{A}$, solving Jónsson's problem. Unfortunately, a year later Herrmann found a gap in his argument which he has not been able to bridge. So the current status of the $\mathrm{CON}\,\mathbf{G} = \mathrm{CON}\,\mathbf{A}$ problem is open.

Finite basis problems and questions of Burnside type. Tarski's problem of finite bases for finite algebras has probably generated more interesting work, while remaining unsolved, than any other problem in the domain of universal algebra. In 1966, he asked, does there exist an algorithm which, given a finite algebra with finitely many basic operations, will determine if the algebra has a finite base for its equations, or its quasi-equations? [Note that by "equation" I mean the same thing as what many writers call an "identity". By an "equational base" for an algebra, I mean a set of equations constituting a complete axiom system for the variety it generates, or what amounts to the same thing, a set of equations valid in the algebra from which all equations valid in it are implied as logical consequences.]

It goes against intuition to suppose that a finite algebra with finitely many operations could ever fail to have a finite equational base, but R. Lyndon in 1954 had produced a seven-element groupoid that has no finite base and by 1966, V. L. Murskii had found a three-element groupoid, and P. Perkins a six-element semigroup, both possessing this property. (Perkins' six-element, nonfinitely based, semigroup consists of the two-by-two $\{0,1\}$-matrices in which 1 occurs at most once, together with the identity matrix, under matrix multiplication.)

On the positive side, S. Oates (now S. Williams) and M. B. Powell had shown in 1965 that every finite group has a finite equational base, and this was nontrivial to prove. By 1973, R. Kruse and I. V. L'vov had independently shown that every finite associative ring is finitely based; and in 1977, K. Baker published his proof (discovered before 1970) that every finite algebra generating a congruence-distributive variety is finitely based. Also, in 1976, S. V. Polin published an example of a nonfinitely based finite non-associative ring, destroying a short-lived conjecture that K. Baker's theorem could be extended to congruence-modular varieties. In 1979, V. L. Murskii published a proof that in a stochastical sense, "almost all" finite algebras are finitely based. Then in 1982, R. M. Bryant confounded everyone with his discovery of a finite nonfinitely based algebra which is simply a group with one distinguished element (or constant operation) adjoined.

Despite all the mentioned results, Tarski's finite basis problem for equations of finite algebras seems more challenging than ever. Nobody is willing to conjecture that it is true, or false; and we are lacking any kind of heuristic argument that could persuade us that the class of finitely based finite algebras is, or is not, likely to be recursively enumerable, or co-recursively enumerable. G. McNulty and C. Shallon [1983] provides a readable survey of the work on this problem that appeared before 1983.

The status of finite basis problems for the quasi-equations satisfied by a finite algebra is similar to that for equations. This area has become more lively since theoretical computer scientists became interested in the problems. Here we are asking whether the quasi-variety generated by a finite algebra $\mathbf{F}$, i.e., the class of all algebras embeddable into direct powers of $\mathbf{F}$, is finitely axiomatizable. Curiously, it turned out that a finite group or associative ring has a finite base of quasi-equations if and only if the variety it generates is residually small. This was proved for groups by A. Yu. Ol'shanskii [1974], and for associative rings by V. P. Belkin [1978], with a very similar argument. Why it should be true remains a mystery. D. Pigozzi [1990] provides a readable introduction to the literature on finite algebras whose quasi-equations are finitely based.

In the late 1970's, the concept of an inherently nonfinitely based finite algebra emerged in the work of V. L. Murskii and P. Perkins. A finite algebra is *inherently nonfinitely based* just in case no finite set of equations valid in it defines a locally finite variety. Since the variety generated by any finite algebra is locally finite, an inherently nonfinitely based finite algebra is nonfinitely based. Moreover, any finite algebra generating a variety that contains an inherently nonfinitely based finite algebra is itself inherently nonfinitely based: the inherent nonfinite basis property is contagious. V. L. Murskii's three-element groupoid and P. Perkins' six-element semigroup have turned out to be inherently nonfinitely based. Of course, R. M. Bryant's group with a constant cannot be.

It looks as though determining the inherently nonfinitely based finite algebras may be a lot easier than determining the nonfinitely based finite algebras; and results in this direction could conceivably suggest an effective approach to Tarski's problems. Therefore, most work on the finite basis problem for equations since about 1980 has been devoted to the determination of which finite algebras are inherently nonfinitely based. G. McNulty, C. Shallon [1983] is also a good introduction to this work.

M. V. Sapir in [1988a] and [1988b] has shown in a most spectacular fashion that the class of inherently nonfinitely based finite semigroups is recursive, and that whether a finite semigroup belongs to this class is mainly determined by the structure of its subgroups. Let $x_1, x_2, \ldots$ be letters (generators of a free semigroup) and inductively define words (elements of the free semigroup) $Z_1 = x_1, Z_{n+1} = Z_n x_{n+1} Z_n$. Sapir's paper [1988b] contains, among other things, the fact that a finite semigroup $\mathbf{S}$ is inherently nonfinitely based if and only if for every n, any equation of the form $W \approx Z_n$ that is valid in $\mathbf{S}$ has W identically equal to Z_n. In [1988a], he reduces the question "does the finite set Σ of semigroup equations define a variety in which all periodic semigroups are locally finite?" to the corresponding question for finite sets of group equations. These results of Sapir constitute a major breakthrough; and I expect that anyone interested in Burnside type questions for groups will find these papers highly stimulating.

Applied universal algebra. We universal algebraists have wished to discover and explore new domains of mathematics, and we believe that we have found some fascinating ones. Like most pure mathematicians, we also have a naive faith that our work is bound to produce results that will eventually prove to be of use in the world outside our field. In the next few paragraphs, I mention some ways in which universal algebra is currently having an impact in the world.

In West Germany, R. Wille has had measurable success in alerting researchers in medicine and a broad range of scientific disciplines to the potential in a form of data-analysis based on elementary lattice theory and the notion of a general Galois connection, which he calls "concept analysis". In the United States, there is a growing dialogue between universal algebra and theoretical computer science. And I am pleased to be able to report that in recent years, universal algebraists have begun to contribute to the solution of problems in branches of pure mathematics besides their own. I will now give two examples of this.

The paper by H. Gross, C. Herrmann, and R. Moresi [1987] furnishes a genuine example of a hard and longstanding open problem of "algebra" whose solution required the involvement of universal algebraists and some techniques that were originated and developed by them. These authors completed the classification of all subspaces of finite dimensional Hermitean vector spaces, up to isometry. They found that underlying Witt's classical work that solved the classification problem in the case of ground fields of characteristic $\neq 2$, there is a fundamental lattice, of six elements, that had always been ignored. In the characteristic 2 case, this lattice explodes into an enormous finite lattice of more than 6,000 elements. Solving the classification problem required determining the structure of this lattice, which in turn required the classification of the 4-generated simple modular lattices (a task initiated by Gel'fand and Ponomarev), as well as more recent investigations of modular polarity lattices carried out principally by C. Herrmann.

In this work of Gross, Herrmann and Moresi we see, I believe, the fruition of a theoretical development that originated in group theory and achieved its fullest development in or on the boundary of universal algebra—viz., the modern theory of modular lattices. After R. Dedekind defined the concept of a modular lattice around 1900 and proved that the lattice of normal subgroups of a group is modular, the first non-trivial results about modular lattices were obtained as direct generalizations of elementary theorems of group theory such as the Jordan-Hölder theorem. Modular lattices turn out to be a difficult subject, but their study has yielded a rich harvest of beautiful results in the past twenty years, achieved by a very few people who mastered their subtleties. A comprehensive account of these developments, which is yet to be written, would include the aforementioned work of R. Gross, C. Herrmann and R. Moresi, the undecidability of the word problem for free modular lattices proved by R. Freese and C. Herrmann, the result of C. Herrmann that the variety generated by modular lattices of finite height is distinct from the variety of all modular lattices, and many results in a similar vein.

My last example involves a theory to which I made several contributions beginning in my years as a graduate student. Since groups are significantly involved, I will devote a few paragraphs to this story. Consider the class of algebras having only one fundamental operation of two variables, the so-called *groupoids*. Let K denote the class of all groupoids, and K_f denote the class of finite groupoids. Thus K_f includes all finite groups and semigroups. Since we can form the Cartesian product of two algebras, we have a binary operation, $\times$, on K_f. The relation of isomorphism, $\cong$, is a congruence relation for $\times$ (on K or on K_f), and so we have a quotient algebra $\langle K_f, \times \rangle / \cong$, which one may call the *algebra of isomorphism types of finite groupoids under direct product*. Letting G_f denote the subclass of finite groups, we also have the algebra $\langle G_f, \times \rangle / \cong$ of isomorphism types of finite groups under direct product. All of these algebras are commutative monoids. It was known to J. H. M. Wedderburn and R. Remak early in this century that finite groups under direct product have unique factorizations, up to isomorphism. In other words, $\langle G_f, \times \rangle / \cong$ is isomorphic to the monoid of positive integers under multiplication, with the isomorphism types of the directly indecomposable finite groups corresponding to the prime integers.

Finite groupoids under direct product are not so well-behaved as finite groups; but the monoid of isomorphism types of finite groupoids does have some prime elements. Tarski had asked whether $\langle K_f, \times \rangle / \cong$ or $\langle K, \times \rangle / \cong$ have any primes; and in my second published paper, R. McKenzie [1968], I showed that the latter monoid has no primes, while every finite, directly indecomposable group represents a prime element among the finite groupoids under direct product. Incidentally, it is still an open problem whether there is any finite algebra (i.e., universal algebra) having two or more basic operations of a positive number of variables which represents a prime element in the monoid of finite algebras with a similar type of operations. A finite group construed as an algebra $\langle G, \cdot, ^{-1} \rangle$ cannot be prime in this sense.

The possibility of extending the unique factorization results proved for finite groups to other kinds of algebras and relational systems has been considered by many authors. Beautiful results of this kind were obtained by Bjarni Jónsson who, aided by his thesis adviser Alfred Tarski, in 1949 proved unique factorization for those finite algebras having a distinguished element 0 such that {0} is a subalgebra, and having among their operations a binary operation $+$ such that $x + 0 = 0 + x = 0$ holds for all elements x of the algebra. We like to call algebras of this kind, finite or infinite, Jónsson-Tarski algebras. Jónsson and Tarski defined a certain concept of *the center* for such algebras, and showed that unique factorization holds for every Jónsson-Tarski algebra satisfying the descending chain condition on subalgebras of the center.

Following in a direct line from this early work of Jónsson and Tarski, there was a series of papers published in the 1960's and early 1970's, typified by C. C. Chang, B. Jónsson and A. Tarski [1964] and R. McKenzie [1971], in which unique factorization was extended to mixed algebraic-relational systems satisfying various finiteness conditions and other restrictions. Twenty years later, some of these results are finding surprising applications in the study of tensor product decompositions of operator algebras. S. C. Power, [1989] and [1990], associates to an operator algebra an invariant which is a binary relational structure, in such a way that the invariant structure associated to the tensor product of operator algebras is the direct product of the structures associated with the factors. He then manages to apply results from the above-mentioned papers to draw conclusions about operator algebras.

Conclusion. I have tried to show that universal algebra has close ties to group theory, and that ideas originating in the two fields interact in numerous, and sometimes mysterious, ways. I omitted from my narrative a number of episodes in which I have observed such interactions. Among others, they involved relation algebras representable as double coset algebras of groups (my PhD thesis); some novel finitely presented groups with unsolvable word problems that were constructed by myself and Richard Thompson about 1968, and that later were involved in the first constructions of unsolvable word problems for lattice-ordered groups; the possibility of giving a definition of the "points" in the first-order language of groups applied to the infinite symmetric groups and, subsequently, a proof by S. Shelah that the "language of elements and permutations" is one of only four true second-order languages; my very minor involvement with the still-open Tarski problems regarding the first-order theories of free groups; and the continuing interest of universal algebraists as well as logicians in the possible existence of nonstandard infinite Jónsson groups, which contributed, I think, to the eventual construction by A. Yu. Ol'shanskii and E. Rips of those fascinating Tarski monster groups.

REFERENCES

[1] Belkin, V. P., 'Quasi-identities of finite rings and lattices', *Algebra i Logika* **17** (1978), 247–259.

[2] Birkhoff, G., 'On the combination of subalgebras', *Proc. Cambridge Philos. Soc.* **29** (1933), 441–464.

[3] Birkhoff, G., 'On the structure of abstract algebras', *Proc. Cambridge Philos. Soc.* **31** (1935), 433–454.

[4] Birkhoff, G., *Lattice Theory*, Third Edition, Colloquium Publications **25** (Amer. Math. Soc., Providence, 1967).

[5] Bryant, R. M., 'The laws of finite pointed groups', *Bull. London Math. Soc.* **14** (1982), 119–123.

[6] Burris, S., Sankappanavar, H. P., *A Course in Universal Algebra*, Graduate Texts in Mathematics **78** (Springer-Verlag, New York, 1981).

[7] Chang, C. C., Jónsson, B., Tarski, A., 'Refinement properties for relational structures', *Fund. Math.* **55** (1964), 249–281.

[8] Cohn, P. M., *Universal Algebra*, Revised Edition (D. Reidel Publ. Co., Boston Dordrecht London, 1981).

[9] Crawley, P., Dilworth, R. P., *Algebraic Theory of Lattices* (Prentice-Hall, Englewood Cliffs, 1973).

[10] Freese, R., 'Free modular lattices', *Trans. Amer. Math. Soc.* **261** (1980), 81–91.

[11] Freese, R., McKenzie, R., 'Residually small varieties with modular congruence lattices', *Trans. Amer. Math. Soc.* **264** (1981), 419–430.

[12] Freese, R., McKenzie, R., *Commutator Theory for Congruence Modular Varieties*, London Math. Soc. Lecture Note Ser. **125** (Cambridge University Press, Cambridge, 1987).

[13] Golubov, E. A., Sapir, M. V., 'Varieties of finitely approximable semigroups', *Soviet Math. Dokl.* **20** (1979), 828–832.

[14] Grätzer, G., *General Lattice Theory* (Academic Press, New York, 1978).

[15] Grätzer, G., *Universal Algebra*, Second Edition (Springer-Verlag, New York, 1979).

[16] Gross, H., Herrmann, C., Moresi, R., 'The classification of subspaces in Hermitean vector spaces', *J. Algebra* **105** (1987), 516–541.

[17] Hagemann, J., Herrmann, C., 'A concrete ideal multiplication for algebraic systems and its relation to congruence-distributivity', *Arch. Math. (Basel)* **32** (1979), 234–245.

[18] Herrmann, C., 'On the word problem for the modular lattice with four free generators', *Math. Ann.* **256** (1983), 513–527.

[19] Hobby, D., McKenzie, R., *The Structure of Finite Algebras*, Contemporary Math. **76** (Amer. Math. Soc., Providence, 1988).

[20] Kurosh, A. G., *Lectures on General Algebra* (Chelsea Publ. Co., New York, 1965).

[21] Lampe, W., 'A property of lattices of equational theories', *Algebra Universalis* **23** (1986), 61–69.

[22] Mal'cev, A. I., *Algebraic Systems* (Springer-Verlag, New York, 1973).

[23] McKenzie, R., 'On finite groupoids and *K*-prime algebras', *Trans. Amer. Math. Soc.* **133** (1968), 115–129.

[24] McKenzie, R., 'Cardinal multiplication of structures with a reflexive relation', *Fund. Math.* **70** (1971), 59–101.

[25] McKenzie, R., 'Residually small varieties of semigroups', *Algebra Universalis* **13** (1981), 171–201.

[26] McKenzie, R., 'Narrowness implies uniformity', *Algebra Universalis* **15** (1982a), 67–85.

[27] McKenzie, R., 'Residually small varieties of K-algebras', *Algebra Universalis* **14** (1982b), 181–196.

[28] McKenzie, R., 'A note on residually small varieties of semigroups', *Algebra Universalis* **17** (1983), 143–149.

[29] McKenzie, R. N., McNulty, G. F., Taylor, W. F., *Algebras, Lattices, Varieties, vol. I* (Wadsworth and Brooks/Cole, Monterey, California, 1987).

[30] McKenzie, R., Valeriote, M., *The Structure of Decidable Locally Finite Varieties*, Progress in Math. **79** (Birkhauser, Boston, 1989).

[31] McNulty, G., Shallon, C., 'Inherently nonfinitely based finite algebras', in *Universal Algebra and Lattice Theory*, ed. by R. Freese and O. Garcia, Lecture Notes in Math. **1004**, pp. 205–231 (Springer-Verlag, Berlin, 1983).

[32] Ol'shanskii, A. Yu., 'Varieties of finitely approximable groups', *Math. USSR—Izv.* **3** (1969), 867–877 (1971).

[33] Ol'shanskii, A. Yu., 'Conditional identities in finite groups', *Siber. Math. J.* **15** (1974), 1000–1003 (1975).

[34] Pálfy, P. P., Pudlák, P., 'Congruence lattices of finite algebras and intervals in subgroup lattices of finite groups', *Algebra Universalis* **11** (1980), 22–27.

[35] Pigozzi, D., 'Finite basis theorems for relatively congruence-distributive quasivarieties', *Proc. Amer. Math. Soc.* (to appear).

[36] Power, S. C., 'Infinite tensor products of upper triangular matrix algebras', *Math. Scand.* (1989) (to appear).

[37] Power, S. C., 'Classifications of tensor products of operator algebras' (preprint).

[38] Sapir, M. V., 'Problems of Burnside type and the finite basis property in varieties of semigroups', *Math. USSR—Izv.* **30** (1988a), 295–314.

[39] Sapir, M. V., 'Inherently nonfinitely based finite semigroups', *Math. USSR—Sb.* **61** (1988b), 155–166.

[40] Sapir, M. V., Shevrin, L. N., 'Residually small varieties of groups and semigroups', *Izv. Vyssh. Uchebn. Zaved. Mat.* (1988), No.10 (317), 41–49.

[41] Smith, J. D. H., *Mal'cev Varieties*, Lecture Notes in Math. **554** (Springer-Verlag, Berlin, 1976).

[42] Taylor, W., 'Residually small varieties', *Algebra Universalis* **2** (1972), 33–53.

Department of Mathematics   and   Department of Mathematics
University of California   La Trobe University
BERKELEY, CA 94720   BUNDOORA, Vic 3083
USA   Australia

GROUPS OF PRIME-POWER ORDER

This lecture is dedicated to Bernhard Neumann.
I have learned many things from him;
for example, the importance of well-chosen examples.

M. F. NEWMAN

1. INTRODUCTION

The most quoted paper on groups is the famous paper of Philip Hall on groups of prime-power order which appeared in the early thirties (1933). This is the foundation stone for a theory of such groups and deserves to be reread from time to time. Today however I want to take up, in part at least, a theme from another important paper of his on p-groups. The paper appeared in Crelle's Journal in 1940 and is an account of the first of a series of lectures by Hall which were at the centre of a Vortragswoche über Gruppentheorie held in Göttingen just over 50 years ago at the end of June 1939. Its title is: The classification of prime-power groups. It was set in the context of determining all groups of a given order. Hall opens:

> The problem of determining all the groups of a given order is an old one, which goes back right to the earliest days of group theory, and was, I believe, initiated by Cayley. It is about this problem as a central theme that I shall develop the remarks which follow.
>
> In order to "determine", that is, to construct all groups of a given kind, it is necessary first to know something of the structure of the groups in question.

Here the intention is construction without repetition. He goes on to talk about some structure theorems which had been used to classify groups of order dividing p^5 and then says:

> These methods, admirable as they are for the purpose for which they were devised, would clearly need to be supplemented, as one passed to groups of higher and higher orders, owing to the gradually increasing complexity of the groups concerned, if equally successful results were to be

This paper is based on a lecture given at the conference. I acknowledge, with gratitude, the helpful comments of Drs Kovács and O'Brien on a draft of this paper.

obtained. And it is natural to ask whether it would not be possible to introduce a system of classification which would apply without modification
to all groups of prime-power order.

It is a *systematic* classification theory of this sort which will be developed in the present article. This theory has grown out of discussions which
I have had over a considerable period with Mr. J. K. Senior of Chicago, to
whom I should like to take this opportunity of expressing my deep indebt-
edness. It is our intention to publish a more detailed account of our results
elsewhere, bearing particularly on the groups of order a power of 2.

Those results eventually appeared, in significant part at least, in 1964 as the tables of Hall and Senior on groups of order dividing 64; so this year is also the silver
anniversary of their publication. I should remind you that the Hall was Marshall Hall
who took over from Philip Hall the final preparation of the tables, and perhaps point
out that Senior was primarily a chemist.

These tables were perhaps the first example of extensive tables of groups. They
are a mine of information and a marvel of accuracy (apart from the permutation representations, see John McKay (1969); these look like an afterthought). In the preface
they say:

It is the hope of the authors that the following tables, ... , will be of
enduring value to those interested in finite groups. Theories change, but
the groups remain.

No single presentation of a group or list of groups can be expected to
yield all the information which a reader might desire. Here each group is
presented in three different ways; ... .

For each group additional information is given.

The count of the number of isomorphism types known then is summarised in the
first six lines of the following table.

Order	Number	First determination
2	1	Cayley (1854)
4	2	Cayley (1854)
8	5	Cayley (1859)
16	14	Hölder (1893), Young (1893)
32	51	Miller (1896)
64	267	Hall, Senior (193?, 1964)
128	2328	James, Newman, O'Brien (1990)
256	56092	O'Brien (1991)

Table 1

Hall proposed classification by what he termed *isoclinism*. I will not remind you
of the formal definition but simply recall that a primary invariant of a group that

is used in this classification is its central quotient, or equivalently its group of inner automorphisms. So that, for example, all abelian groups have the same isoclinism type.

This paper is a report on some recent work which involves somewhat different classification. It is, in a sense, a report on progress on a research program I outlined in a lecture given here in Canberra in 1975 (see Newman (1977)). At that time it was not at all clear how computers would impact on the study of p-groups. The impact has been considerable. There have been further determinations of all groups of a given order such as those of order 2^7 and 2^8 given in Table 1. Some information on these groups and the information in the Hall-Senior tables has been made available in a more usable form (see Newman and O'Brien (1989)). This is available in standalone form and as libraries within the system CAYLEY (see Cannon 1984) and is expected to become available in the relatively new, Aachen-based, system GAP (see Schönert (1989)). Note that in the context of such systems it is no longer necessary to have available more than one description of a group or to give additional information on the group. It may, of course, be desirable to store the results of long calculations, and other material such as bibliographic information. In another direction, the study of p-groups of large nilpotency class has been opened up and striking progress made. There has also been progress on the study of Burnside groups of prime-power exponent but I will not mention that further here (see Vaughan-Lee (1990)). An important part of the progress comes from the ability to keep track of large amounts of information and to avoid errors in calculation; quite a number of such errors have been revealed and some more precise warnings about them will be given later.

The classifications can be viewed as using various directed graph structures on the set of isomorphism types of p-groups. They are described in Section 2.

Before I go into more detail let me say something about the notion that things get more complicated as the order or composition length increases. This might be called, borrowing from the title of a history of Australia by Blainey (1966), the tyranny of the large. While this tyranny is real enough it is not the whole story. For example Hall's work (1933) shows that, given a positive integer n, for all primes p with $p \geqslant n$ the groups of order p^n behave in a regular way because the interaction between the power and the commutator structure is reasonably good. Thus it is the **small** primes which cause additional trouble. So there might also be said to be a tyranny of the small. This is overcome to some extent by the computer programs which allow the study of p-groups for a fixed prime p. The work on groups of large nilpotency class shows that the behaviour of these groups settles down once the composition length is sufficiently large. (The appropriate term 'settled' comes from a manuscript of Leedham-Green.) This observation provides another example of the tyranny of the small; a specific instance will be given in the story of 5-groups of maximal nilpotency class in Section 4. This tyranny of the small in some ways resembles what Guy (1988) has called the strong law of small numbers; of his formulations perhaps the relevant one here is "Early exceptions eclipse eventual essentials".

In Section 3 I will make some remarks about groups of small composition length and I will close (Section 5) with an attempt at a more precise form for Cayley's general problem (1878) of listing finite groups.

2. DIRECTED GRAPHS

In this section three kinds of directed graph (digraph) structure are described. For each prime p, the underlying set is always a set S_p of groups of p-power order such that each finite p-group is isomorphic to a group in S_p and no two groups in S_p are isomorphic.

The first kind of digraph, U_p, is implicit in the paper (1940) of Hall. Let $\zeta(G)$ denote the centre of a group G. The edges of U_p are the pairs (P, Q) with P isomorphic to the central quotient $Q/\zeta(Q)$. Let us say in this case that Q is an *immediate descendant* of P and say that R is a *descendant* of P if there is a (possibly empty) path from P to R in U_p. In these terms all p-groups are descendants of the identity group E and all abelian p-groups are immediate descendants of E. The latter form the first isoclinism family, Φ_1, in Hall's classification. A group is said to be *capable* if it has immediate descendants. Hall (1940, p. 137) discusses conditions for a group to be capable. For example, he shows that a necessary condition for a p-group of nilpotency class less than p to be capable is that the top two type-invariants be equal. Thus the non-trivial cyclic groups are not capable (or are *terminal*). On the other hand all the non-cyclic elementary abelian groups are capable. The immediate descendants of the elementary abelian group of order p^2 are the non-abelian groups with centre of index p^2; these groups all have commutator subgroup of order p (though not conversely); they form Hall's second isoclinism family, Φ_2. Hall showed that there are 10 isoclinism families containing groups of order up to p^5 (for $p \geqslant 5$) and Easterfield (1940) extended this to p^6 showing that there are then 43 isoclinism families. The recent work of Wilkinson (1988) shows that the number of isoclinism families containing groups of order p^7 depends on p.

The definition of the second family of digraphs, $\mathcal{L}_p$, uses the lower central series. It will be convenient to use the following notation. The commutator of two subgroups, A and B, of G will be denoted $[A, B]$. The lower central series of G is

$$G = \gamma_1(G) \geqslant \cdots \geqslant \gamma_i(G) \geqslant \cdots$$

where $\gamma_{i+1}(G) = [\gamma_i(G), G]$. The edges of $\mathcal{L}_p$ are the pairs (P, Q) with P isomorphic to the quotient $Q/\gamma_c(Q)$ where $\gamma_c(Q)$ is the last non-trivial term of the lower central series of Q. The first time I saw such a digraph used was in a lecture by Leedham-Green at the International Congress of Mathematicians in Vancouver in 1974. In these digraphs the immediate descendants of the elementary abelian group of order p^2 are the two non-abelian groups of order p^3. The digraphs $\mathcal{L}_p$ have the advantage over the $\mathcal{U}_p$ that every non-trivial group in $\mathcal{L}_p$ has only finitely many immediate descendants. Moreover all the immediate descendants of a group P in $\mathcal{L}_p$ have the same generator number as P. A result of Taussky (1937) shows that in $\mathcal{L}_2$ the descendants of the elementary abelian group of order 4 are the dihedral, semidihedral and generalised quaternion groups. The subdigraph of $\mathcal{L}_2$ of all descendants of the elementary abelian group of order 4 is easy to picture (it is drawn in Leedham-Green and Newman (1980)).

The digraphs $\mathcal{L}_p$ were the basis of some computations by Leedham-Green and others that I will mention later. For computational, and also some theoretical, purposes it is more convenient to use the lower exponent-p central series, for which I use the notation:

$$G = \mathcal{P}_0(G) \geqslant \cdots \geqslant \mathcal{P}_i(G) \geqslant \cdots$$

where $\mathcal{P}_{i+1}(G) = [\mathcal{P}_i(G), G]\mathcal{P}_i(G)^p$. (The usual mismatch of numbering between these two central series is retained; the corresponding *classes* are distinguished here by always referring to the former as *nilpotency* class). This view leads to a third family of digraphs, $\mathcal{K}_p$, whose edges are the pairs (P, Q) with P isomorphic to the quotient $Q/\mathcal{P}_c(Q)$ where $\mathcal{P}_c(Q)$ is the last non-trivial term of the lower exponent-p central series of Q. In this case the elementary abelian group of order p^2 has 7 immediate descendants: 3 of order p^3, 3 of order p^4 and 1 of order p^5.

Parts of a digraph can be described in tabular form. As an example the following table describes the subdigraph of $\mathcal{K}_5$ consisting of groups of order 5^n and nilpotency class $n - 1$ for $n \leqslant 8$. Note that in this subdigraph the order of the group at the head of an edge is always 5 times that of the group at the tail of the edge. Let me explain the meaning of the entries in the middle section of the table by considering the second row. There are *three* entries which indicate that the $3 = 2 + 1$ capable groups of order 5^6 have, respectively, 12 immediate descendants of which 2 are capable, 17 immediate descendants of which 2 are capable and 70 immediate descendants of which 1 is capable. The last row consists of the roots. A * indicates that the capable groups concerned have further immediate descendants in $\mathcal{K}_5$.

Order	Immediate descendants / capable					Total
5^8	14 / 2	18 / 2	49 / 1	37 / 1	33 / 0	151
5^7	12 / 2	17 / 2	70 / 1			99
5^6	21 / 2	18 / 1				39
5^5	9 / 2*					9
5^4	4 / 1					4
5^3	2 / 1*					2
5^2	2 / 1*					2

Table 2

Relative to such a digraph the groups can be classified recursively by their position in the digraph. The primary invariant of a group is its *parent*, the group of which it is an immediate descendant. The immediate descendants Q of a group P can then be further classified by their *step size* which is $\log_p(|Q|/|P|)$. It will require further study before it makes much sense to propose a meaningful classification for the immediate descendants with a fixed step size. From now on, unless otherwise indicated, the digraphs under consideration will be the $\mathcal{K}_p$.

The crux of the point of view presented here is that these digraphs can be explored by calculating the set of immediate descendants of a vertex. An outline for an algorithm for doing this was described in my 1975 lecture mentioned earlier. (I have recently discovered that Graham Higman had briefly outlined a similar idea in a seminar in Chicago in 1960(c).) The algorithm was partially implemented in 1976 and the implementation completed in 1986. A detailed account by O'Brien of the algorithm and its implementation will appear soon (1990).

An early exploration with this program was a study of 3-groups of order 3^n and nilpotency class $n-2$ for $n \leqslant 10$ by Ascione, Havas and Leedham-Green (1977). As well as the other cases already mentioned it has been used to determine the 505 groups of order 3^6 (Baldwin, Newman, O'Brien and Ozols, unpublished). It was used by Wilkinson (1988) to guide his study of groups of order p^7 and exponent p (he should perhaps have done more such investigations because, as Tyler has pointed out (see the review in MR), some errors have crept in). It has also been used to explore the subdigraph of metacyclic p-groups for $p = 2, 3$ (see Newman and Xu (1988)).

It is natural to ask how far one can go with such computer calculations. For example can one determine the groups of order 2^9? It would need about 13000 immediate descendant calculations almost all of which are straightforward. But there are a few delicate ones such as the calculation of the immediate descendants of step size 3 of the elementary abelian group of order 64. Moreover lower bounds by Higman (1960a) show that there are at least 8.4 million isomorphism types of groups of this order so storage and retrieval methods will need to be significantly improved for the results of such a calculation to be made available for use.

Useful progress can also be made by hand, at times using computer generated results for guidance. For instance one can calculate the immediate descendants of certain parametrised families of groups. An example that has already been mentioned above is the family of elementary abelian groups of order p^2 parametrised by the prime p; though it should be noted that the calculation for the prime 2 differs in some details from that for odd primes. More extensive calculations of this sort were done by Küpper in her Master's thesis (1979) in which she determined the immediate descendants of all the (26) capable 2-generator p-groups of order up to p^5 for $p \geqslant 7$. Another example is the family of dihedral groups of order 2^n parametrised by the integer n; each has 3 immediate descendants: the dihedral, semidihedral and generalised quaternion groups of order 2^{n+1}. A similar example for a family of 3-groups has been studied by Ascione (1979). The cyclic groups of order p^n parametrised by p and n can clearly be handled in one immediate descendant calculation to show that they each have one immediate descendant, the cyclic group of order p^{n+1}. A somewhat more elaborate calculation of this kind has been used by Newman and Xu (in preparation) to determine the subdigraph of $\mathcal{K}_p$ consisting of all metacyclic p-groups.

3. SMALL COMPOSITION LENGTH

The groups of order p^3 are very well known. All the immediate descendant calculations needed to determine them in $\mathcal{K}_p$ have already been mentioned. For odd primes the distinct groups can be given by presentations of the form

$$\{\, a_1, a_2, a_3 : [a_2, a_1] = a_3^\alpha,\ [a_3, a_1] = [a_3, a_2] = \emptyset,$$
$$a_1^p = a_2^\beta a_3^\gamma,\ a_2^p = a_3^\delta,\ a_3^p = \emptyset \,\}$$

where $\alpha, \beta, \gamma, \delta$ are given by the following table:

	α	β	γ	δ	Description	
1	0	1	0	1	cyclic p^3	capable
2	1	0	0	0	non-abelian, exp p	capable
3	1	0	1	0	non-abelian, exp p^2	terminal
4	0	0	1	0	abelian (p^2, p)	capable
5	0	0	0	0	abelian (p, p, p)	capable

Table 3

It would be possible to make this table apply for all primes by altering the value of δ in row 3 but this would mask the fact, mentioned earlier, that some details of the proof for the prime 2 are slightly different.

The groups of order p^4 are well-known. They are treated, usually in an ad hoc manner, in quite a number of texts from Burnside (1897, p. 82) to Suzuki (1986, p. 85). The calculation of the immediate descendants of step size 1 for the 4 capable groups of order p^3 gives 11 groups of which 5 are capable (1/1, 4/1, 2/1, 4/2 respectively). Adding to these the 3 immediate descendants of step size 2 of the elementary abelian group of order p^2, all of which are capable, and the elementary abelian group of order p^4 gives the 15 groups of which 9 are capable.

The first determination of groups of order p^5 was given by Bagnera in 1898. These groups can not be described as well known. There is an account in de Séguier's book (1904, p. 134) and Schreier (1926) used them to exemplify his work on the theory of

extensions. There is a recent list of presentations in James' paper (1980). The number
of groups is $2p + 2\gcd(p-1,3) + \gcd(p-1,4) + 61$ for $p \geqslant 5$. Hence it is clearly
no longer possible to list all the groups of order p^5 by a finite set of presentations
parametrised by p alone. This can be overcome by introducing further parameters.
What is important is to do this in a way which reflects the structure of the groups in
question. This dependence of the number of groups on the prime arises in considering
the immediate descendants of step size 2 for two of the groups of order p^3, namely the
non-abelian capable group and the elementary abelian group. Take as an example the
non-abelian group. It has $p + 7$ immediate descendants of step size 2 (of which 6 are
capable). These can be classified further by the order of the subgroup of p-th powers.
In $p + 3$ of these immediate descendants this subgroup has order p^2. These can now be
further classified using the automorphism groups. Let me omit the details. Finally one
ends up with a few individual groups and the following two families of presentations for
pairwise non-isomorphic groups.

$$\{\, a_1, a_2, a_3, a_4, a_5 : a_3 = [a_2, a_1], \; a_4 = [a_3, a_1], \; a_5 = [a_3, a_2],$$
$$[a_4, a_1] = [a_4, a_2] = [a_5, a_1] = [a_5, a_2] = \emptyset,$$
$$a_1^p = a_5^{\rho-1}, \; a_2^p = a_4 a_5^2,$$
$$a_3^p = a_4^p = a_5^p = \emptyset \,\}$$

where ρ runs through the set of the $(p-1)/2$ non-squares modulo p between 0 and p;

$$\{\, a_1, a_2, a_3, a_4, a_5 : a_3 = [a_2, a_1], \; a_4 = [a_3, a_1], \; a_5 = [a_3, a_2],$$
$$[a_4, a_1] = [a_4, a_2] = [a_5, a_1] = [a_5, a_2] = \emptyset,$$
$$a_1^p = a_4^\sigma, \; a_2^p = a_5,$$
$$a_3^p = a_4^p = a_5^p = \emptyset \,\}$$

where σ runs through a set of $(p-3)/2$ representatives of the non-trivial cosets of the
subgroup of order 2 in the multiplicative group of non-zero residues modulo p. Within
each family the groups have very similar structures. In summary it turns out that one
can divide the primes into 6 families, the congruence classes modulo 12, so that for all
the primes p in a family it is possible to describe all the groups of order p^5 by a single
set of parametrised presentations.

The story for groups of order p^6 is not yet completely told. It has been the
subject of a number of theses starting with Potron's (1904a). The others I know of are
by Tordella (1939), Easterfield (1940), James (1968), Küpper (1979), and Pilyavskaya
(1983a). Potron (1904b) and Pilyavskaya (1983b) have published brief summaries of
their work. The only reasonably accessible full list of presentations is that published
by James in 1980. While this list is substantially correct it contains some errors. In
fact none of the treatments is error-free. O'Brien and I have a list of some errors and

corrections for them. We hope, in time, to create an electronic list of parametrised presentations which reflects the structure of the groups.

As already mentioned a beginning has been made on the determination of the groups of order p^7. There have been theses by Tyler (1982) and Wilkinson (1983) on the groups of exponent p.

4. NILPOTENCY COCLASS

The work of Leedham-Green and McKay (1976, 1978a,b) on p-groups of maximal (nilpotency) class suggested the use of nilpotency coclass as a primary classification invariant: a group of order p^n and nilpotency class c has *nilpotency coclass* $n-c$. So the groups of maximal class have nilpotency coclass 1. This work also suggested a sequence of conjectures on the structure of groups classified by nilpotency coclass (Leedham-Green and Newman (1980)). Leedham-Green (Oberwolfach, 1987) has announced the confirmation of these conjectures; this results from some deep and interesting work by Leedham-Green and various collaborators. These results show, among other things, that given a prime p and a positive integer r there is a positive integer m such that all descendants (not just the immediate ones) of all groups of order p^m and nilpotency coclass r have nilpotency coclass r; that is, their nilpotency coclass is settled. Several of the corresponding digraphs have been explored by computer. As already mentioned 3-groups of nilpotency coclass 2 up to order 3^{10} were studied by Ascione et al. (1977). This has been taken further first by Ascione (1979) and then by Leedham-Green and Newman (unpublished). More recently Newman and O'Brien (in preparation) explored the digraph for 2-groups of coclass at most 3. (In the process we have found some errors in James' work on 2-groups of nilpotency coclass 2 (1975, 1983)). As an example of such computer calculations I will describe here some of the work that has been done in exploring the subdigraph $\mathcal{C}$ of $\mathcal{K}_5$ consisting of groups with nilpotency coclass 1.

The digraph $\mathcal{C}$ has an isolated point, the cyclic group of order 25, and the rest is a tree rooted at the elementary abelian group of order 25. The first few levels of $\mathcal{C}$ are given by Table 2. The work of Blackburn (1958), which mostly deals with a general prime, yields a significant amount of further information about the structure of $\mathcal{C}$. First, it has a single unbounded path. This consists of the (non-cyclic) finite quotients of the central quotient of the wreath product of the additive group of 5-adic integers by a cyclic group of order 5. Let me denote the quotient of order 5^m by U_m. Second, all the descendants of the groups of order 5^6 in $\mathcal{C}$ have nilpotency coclass 1. So the tree can be viewed as consisting of the subtree of such groups of order at most 5^6 together with the full descendant trees of the capable groups of order 5^6. Third, every group in $\mathcal{C}$ of order 5^n has a quotient U_m with $2m \geqslant n$. Thus every path beginning at U_m and not passing through U_{m+1} has length at most m. Moreover Blackburn mentioned an example of a group of order 5^{14} which is not a descendant of U_8.

Leedham-Green and McKay in the first (1976) of their papers acknowledge the "valuable insights" they gained from "the instant computer calculations of A. Learner". One aspect of this was the work of Maung, a student of Leedham-Green, who used a computer (in about 1975) to show that there are 99 groups of order 5^7 and nilpotency class 6. This program was based on the use of the lower central series. Leedham-Green and Maung extended this to a determination of the 5-groups of maximal class up to order 5^{16} which are not descendants of U_{13}. Leedham-Green and McKay went on (1978a,b) to construct (with repetitions) many 5-groups of nilpotency coclass 1. For each m they constructed, in effect, all descendants of U_m which are not descendants of U_{m+1} and whose order is at most 5^{2m-2} (call them type m) and some of order 5^{2m}. The latter show that Blackburn's upper bound for the order of descendants of U_m is uniformly sharp. They took the theory further and in 1984 were, as a special case of a more general result, able to solve the isomorphism problem for central quotients of groups of type m. These quotients correspond precisely to the capable vertices up to level $m-3$ in the subdigraph described in Table 4(m) below.

In 1980 I took these calculations somewhat further, in places up to groups of order 5^{30}. This revealed what seems to be a doubly periodic structure with some minor irregularities for small orders (up to 5^{16} as it happens). More recently I have repeated the calculation systematically for descendants of the U_m with $m \leqslant 17$. This involved more than 17000 groups and it took about 2 days of computer time on a Microvax 2.

On the basis of this one can attempt an abstract definition of $\mathcal{C}$. Let $a(i,j)$ be given by the (i',j') entry of the following table:

	0	1	2	3
0	9	8	11	8
1	16	17	16	18
2	11	8	9	8
3	16	27	16	29

where i',j' are the residues modulo 4 of i,j. Let $e(1,m) = 9 + 2\gcd(m-1,4)$ and $e(m,m) = 1 + 2\gcd(m+1,2) + 5\gcd(m+1,4)$. Consider the digraphs given by Table 4(m). The computer calculations show that for $m = 9, \ldots, 17$ the tree of descendants of U_m which are not descendants of U_{m+1} has the structure given by that table. It should be noted that this does not hold for $m \leqslant 8$ (except $m = 5$). Nevertheless it is compelling to conjecture that these exceptions are examples of the tyranny of small orders and that the tree of descendants of U_m which are not descendants of U_{m+1} has the structure given by Table 4(m) for all $m \geqslant 9$.

Exponent	Immediate descendants / capable						
$2m$	$e(m,m)$ / 0	25 / 0	25 / 0	25 / 0	25 / 0	25 / 0	25 / 0
$2m-1$	$a(m-1,m)$ / 1	29 / 1	29 / 1	29 / 1	29 / 1	29 / 1	29 / 1
⋮	⋮	⋮	⋮	⋮	⋮	⋮	⋮
n	$a(n-m,m)$ / 1	29 / 1	29 / 1	29 / 1	29 / 1	29 / 1	29 / 1
⋮	⋮	⋮	⋮	⋮	⋮	⋮	⋮
$m+7$	$a(7,m)$ / 1	29 / 1	29 / 1	29 / 1	29 / 1	29 / 1	29 / 1
$m+6$	$29 + a(6,m)$ / 2		145 / 5				
$m+5$	$a(5,m)$ / 1		29 / 1				
$m+4$	$a(4,m)$ / 1		29 / 1				
$m+3$	$10 + a(3,m)$ / 1		49 / 1				
$m+2$	$9 + a(2,m)$ / 2						
$m+1$	$e(1,m)$ / 1						
m	1 / 1						

Table 4(m)

The algorithm involves computing information about the automorphism groups of all the groups involved. These accompanying automorphism calculations show that the automorphism groups of most of these groups are 5-groups. If the conjecture is correct, then over 90% have this property. This goes along the same lines as the asymptotic result of Martin (1986) and does not conflict with it.

There have been some initial explorations of the digraph of 7-groups of nilpotency coclass 1 (see Leedham-Green and McKay (1984) p. 303).

5. A QUESTION

It seems that the tyranny of the small is starting to be broken for p-groups.

The evidence is perhaps beginning to suggest that we should try to find out whether given a positive integer n the set Π of all primes can be **finitely** partitioned

$$\Pi = \Pi_1 \cup \ldots \cup \Pi_{t(n)}$$

so that for each i it is possible to list the groups of order p^n with p in Π_i by a single finite list of parametrised presentations.

An enumerative conjecture along these lines is the PORC conjecture of Higman (1960a,b).

In this connection it may pay to explore the sort of digraph structures I have been describing. It may also be worthwhile to use the ideas of symbolic collection of Leedham-Green and Soicher (1990).

In the light of the "classification" theorem for finite simple groups one might even hope that such a situation prevails for all finite groups. Given n consider the set Σ of all multisets (allows repetitions) of n simple groups. *Can the set Σ be finitely partitioned*

$$\Sigma = \Sigma_1 \cup \ldots \cup \Sigma_{t(n)}$$

so that for each i the groups with composition multiset in Σ_i can be described by a finite set of parametrised descriptions? The classification theorem shows the answer is yes for $n = 1$.

REFERENCES

Judith A. Ascione, George Havas and C. R. Leedham-Green (1977), 'A computer aided classification of certain groups of prime power order', *Bull. Austral. Math. Soc.* **17**, 257–274. Corrigendum: 317–319. Microfiche supplement: 320.

Judith A. Ascione (1979), *On 3-groups of second maximal class*, Ph.D. thesis, Australian National University.

G. Bagnera (1898), 'La composizione dei Gruppi finiti il cui grado è la quinta potenza di un numero primo', *Ann. Mat. Pura Appl.* (3) **1**, 137–228.

N. Blackburn (1958), 'On a special class of p-groups', *Acta. Math.* **100**, 45–92.

Geoffrey Blainey (1966), *The tyranny of distance* (Sun Books, Melbourne; also Macmillan, Melbourne, 1968).

W. Burnside (1897), *Theory of groups of finite order* (Cambridge University Press).

John J. Cannon (1984), 'An introduction to the group theory language, Cayley', in *Computational Group Theory*, Proceedings, Durham, 1982; ed. by Michael D. Atkinson, pp. 145–183 (Academic Press, London, New York).

A. Cayley (1854), 'On the theory of groups, as depending on the symbolic equation $\theta^n = 1$', *Philos. Mag.* (4) **7**, 40–47. Mathematical Papers, II, 123–130.

A. Cayley (1859), 'On the theory of groups, as depending on the symbolic equation $\theta^n = 1$. Part III', *Philos. Mag.* (4) **18**, 34–37. Mathematical Papers, IV, 88–91.

A. Cayley (1878), 'Desiderata and suggestions. No. 1. the theory of groups', *Amer. J. Math.* **1**, 50–52. Mathematical Papers, X, 401–403.

J.-A. de Séguier (1904), *Théorie des groupes finis. Éléments de la théorie des groupes abstraits* (Gauthier-Villars, Paris).

Thomas E. Easterfield (1940), *A classification of groups of order p^6*, Ph.D. thesis, Cambridge University.

Richard K. Guy (1988), 'The strong law of small numbers', *Amer. Math. Monthly* **95**, 697–712.

Marshall Hall, Jr. and James K. Senior (1964), *The Groups of Order 2^n ($n \leqslant 6$)* (Macmillan, New York).

P. Hall (1933), 'A contribution to the theory of groups of prime-power order', *Proc. London Math. Soc.* **36** (1934), 29–95. Collected Works, 59–125.

P. Hall (1940), 'The classification of prime-power groups', *J. Reine Angew. Math.* **182**, 130–141. Collected Works, 265–276.

Graham Higman (1960a), 'Enumerating p-groups. I: Inequalities', *Proc. London Math. Soc.* (3) **10**, 24–30.

Graham Higman (1960b), 'Enumerating p-groups. II: Problems whose solution is PORC', *Proc. London Math. Soc.* (3) **10**, 566–582.

Graham Higman (1960c), 'Enumerating p-groups: III'. One of four lectures delivered as part of a group theory seminar at the University of Chicago in Autumn 1960.

Otto Hölder (1893), 'Die Gruppen der Ordnungen p^3, pq^2, pqr, p^4', *Math. Ann.* **43**, 301–412.

Rodney K. James (1968), *The Groups of Order p^6 ($p \geqslant 3$)*, Ph.D. thesis, University of Sydney.

Rodney James (1975), '2-groups of almost maximal class', *J. Austral. Math. Soc. Ser. A* **19**, 343–357.

Rodney James (1980), 'The groups of order p^6 (p an odd prime)', *Math. Comput.* **34**, 613–637.

Rodney James (1983), '2-groups of almost maximal class: corrigendum', *J. Austral. Math. Soc. Ser. A* **35**, 307.

Rodney James, M. F. Newman and E. A. O'Brien (1990), 'The groups of order 128', *J. Algebra* **129**, 136–158.

A. M. Küpper (1979), *Enumeration of some two-generator groups of prime power order*, M.Sc. thesis, Australian National University.

C. R. Leedham-Green and Susan McKay (1976), 'On p-groups of maximal class I', *Quart. J. Math. Oxford* (2) **27**, 297–311.

C. R. Leedham-Green and Susan McKay (1978a), 'On p-groups of maximal class II', *Quart. J. Math. Oxford* (2) **29**, 175–186.

C. R. Leedham-Green and Susan McKay (1978b), 'On p-groups of maximal class III', *Quart. J. Math. Oxford* (2) **29**, 281–299.

C. R. Leedham-Green and Susan McKay (1984), 'On the classification of p-groups of maximal class', *Quart. J. Math. Oxford* (2) **35**, 293–304.

C. R. Leedham-Green and M. F. Newman (1980), 'Space groups and groups of prime-power order I', *Arch. Math. (Basel)* **35**, 193–202..

C. R. Leedham-Green and L. H. Soicher (1990), 'Collection from the left and other strategies', *J. Symbolic Comput.* (to appear).

Ursula Martin (1986), 'Almost all p-groups have automorphism group a p-group', *Bull. Amer. Math. Soc. (N.S.)* **15**, 78–82.

John McKay (1969), 'Table errata: *The Groups of Order 2^n ($n \leqslant 6$)*, (Macmillan, New York, 1964) by M. Hall and J. K. Senior', *Math. Comput.* **23**, 691–692.

G. A. Miller (1896), 'The regular substitution groups whose order is less than 48', *Quart. J. Math.* **28**, 232–284.

M. F. Newman (1977), 'Determination of groups of prime-power order', in *Group Theory*, Proceedings, Canberra, 1975; ed. by R. A. Bryce, J. Cossey and M. F. Newman; Lecture Notes in Math. **573**, pp. 73–84 (Springer-Verlag, Berlin, Heidelberg, New York).

M. F. Newman and E. A. O'Brien (1989), 'A CAYLEY library for the groups of order dividing 128', in *Group Theory*, Proceedings of the Singapore Group Theory Conference held at the National University of Singapore, 1987; ed. by Kai Nah CHENG and Yu Kiang LEONG, pp. 437–442 (de Gruyter, Berlin, New York).

M. F. Newman and Mingyao Xu (1988), 'Metacyclic groups of prime-power order', *Adv. in Math. (Beijing)* **17**, 106–107.

E. A. O'Brien (1990), 'The p-group generation algorithm', *J. Symbolic Comput.* (to appear).

E. A. O'Brien (1991), 'The groups of order 256', *J. Algebra* (to appear).

O. S. Pilyavskaya (1983a), *Classification of groups of order p^6 ($p > 3$)*, VINITI Deposit No. 1877–83.

O. S. Pilyavskaya (1983b), 'Application of matrix problems to the classification of groups of order p^6, $p > 3$', in *Linear algebra and the theory of representations*, ed. by Yu. A. Mitropol'skiĭ, pp. 86–99 (Inst. Mat. Akad. Nauk Ukrain. SSR, Kiev).

M. Potron (1904a), *Les groupes d'ordre p^6*, doctoral thesis (Gauthier-Villars, Paris).

M. Potron (1904b), 'Sur quelques groupes d'ordre p^6', *Bull. Soc. Math. France* **32**, 296–300.

Martin Schönert (1989), 'GAP, groups and programming', *Abstracts Amer. Math. Soc.* **10**, 34.

Otto Schreier (1926), 'Über die Erweiterung von Gruppen. II', *Abh. Math. Sem. Univ. Hamburg* **4**, 321–346.

Michio Suzuki (1986), *Group Theory II*, Grundlehren Math. Wiss. **248** (Springer-Verlag, Berlin, Heidelberg, New York).

Olga Taussky (1937), 'A remark on the class field tower', *J. London Math. Soc.* **12**, 82–85.

Louis William Tordella (1939), *A classification of groups of order p^6, p an odd prime*, Ph.D. thesis, University of Illinois, Urbana.

Douglas Blaine Tyler (1982), *Determination of groups of exponent p and order p^7*, Ph.D. dissertation, University of California, Los Angeles.

Michael Vaughan-Lee (1990), *The restricted Burnside problem*, London Math. Soc. Monographs **5** (Clarendon Press, Oxford).

J. W. A. Young (1893), 'On the determination of groups whose order is a power of a prime', *Amer. J. Math.* **15**, 124–178.

David Fredric Wilkinson (1983), *An application of computers to groups of prime exponent*, Ph.D. thesis, University of Warwick.

David Wilkinson (1988), 'The groups of exponent p and order p^7 (p any prime)', *J. Algebra* **118**, 109–119.

Mathematics IAS
Australian National University
GPO Box 4
CANBERRA ACT 2601
Australia

FINITE PRIMITIVE PERMUTATION GROUPS: A SURVEY

Dedicated to B. H. Neumann, my mentor and friend
with love and good wishes for his eightieth birthday

CHERYL E. PRAEGER

Over the last ten years remarkable progress has been made in analysing and describing the structure of finite permutation groups, especially finite primitive permutation groups. Perhaps the major reason has been the completion of the classification of the finite simple groups. This paper contains a discussion of some striking new results about finite primitive permutation groups, many of them requiring the finite simple group classification for their proof. Moreover, many newly developed techniques for working with permutation groups have been used to solve problems in areas related to group theory, and we shall discuss some results of this type in the areas of algebraic graph theory, algebraic number theory, the theory of combinatorial designs and of algorithms for permutation group computations. There were several survey articles written about the consequences of the finite simple group classification for permutation groups and related areas soon after the classification was completed, see [22], [68], [69], [97], [110], [111]. Perhaps the most well-known of these is the article of Cameron [22]. This paper has minimal overlap with any of these, either in content or approach. Most of the results surveyed here post-date these articles, and the discussion is aimed at demonstrating a variety of techniques which have been used successfully to apply the finite simple group classification to solve problems about permutation groups.

For a finite set Ω of n points, the *symmetric group on* Ω, that is the group of all permutations of Ω, will be denoted by $\mathrm{Sym}(\Omega)$ or S_n. A *permutation group on* Ω is a subgroup G of $\mathrm{Sym}(\Omega)$; such a subgroup G is also called a *permutation group of degree* n. For each permutation group G on Ω there corresponds an equivalence relation on Ω, where points α, β of Ω are related if and only if $\alpha^g = \beta$ for some $g \in G$. The equivalence classes of this relation are called *orbits* of G, or *G-orbits*, in Ω, and G is said to be *transitive* on Ω if there is just one G-orbit in Ω, that is if $\Omega = \alpha^G := \{ \alpha^g \mid g \in G \}$. For a permutation group G on Ω to be *primitive* it must be transitive and it must satisfy one further condition, namely the only partitions of Ω

The author thanks the Australian National University for its support in the form of a Visiting Fellowship in the School of Mathematical Sciences during the period when this paper was prepared.

it may preserve are the trivial ones. A permutation group G on Ω is said to *preserve* a partition, $P = \{B_1|B_2|\ldots|B_r\}$, of Ω if, for each $g \in G$ and each block B_i of the partition P, the image B_i^g is also a block of the partition, that is G permutes the blocks of the partition blockwise. A partition of Ω is called *trivial* if either the partition consists of a single block $\{\Omega\}$, or each block of the partition has size 1. If a transitive permutation group G on Ω is not primitive it is said to be *imprimitive* on Ω, and is in a sense made up of two transitive permutation groups of smaller degree, namely if G preserves the nontrivial partition $P = \{B_1|\ldots|B_r\}$ then G is a subgroup of the wreath product $K \operatorname{wr} L$ of the transitive group $K \leqslant \operatorname{Sym}(B_1)$ induced by G on a block B_1 of the partition and the transitive group $L \leqslant S_r$ induced by G on the set of blocks. Thus the finite primitive permutation groups are the fundamental finite permutation groups in the sense that they are not made up of smaller degree groups as described above. Moreover every finite permutation group is a subgroup of a wreath product $K_1 \operatorname{wr} K_2 \operatorname{wr} \ldots \operatorname{wr} K_r$ of primitive groups K_1, K_2, $\ldots$, K_r.

1. MULTIPLY TRANSITIVE GROUPS

One of the first consequences of the finite simple group classification for finite permutation groups was the classification of multiply transitive groups which are defined as follows. For a positive integer k, a permutation group $G \leqslant \operatorname{Sym}(\Omega)$ is said to be *k-transitive* on Ω if G is transitive on the set

$$\Omega^{(k)} = \big\{ (\alpha_1, \ldots, \alpha_k) \mid \alpha_i \text{ distinct points of } \Omega \big\}.$$

Thus the 1-transitive groups are the transitive groups, and, if $k \geqslant 2$, it is easy to show that G is k-transitive on Ω if and only if G is transitive on Ω and, for $\alpha \in \Omega$, the stabilizer G_α is $(k-1)$-transitive on $\Omega - \{\alpha\}$. A permutation group G is said to be *multiply transitive* if it is k-transitive for some $k \geqslant 2$. Clearly all multiply transitive groups are primitive. Examples of finite multiply transitive groups include the following.

EXAMPLES 1.1. (a) The symmetric group S_n is n-transitive, $n \geqslant 2$, and the alternating group A_n, the group of all even permutations in S_n, is $(n-2)$-transitive, $n \geqslant 4$.

(b) The Mathieu groups $M_n \leqslant S_n$, $n \in \{11, 12, 22, 23, 24\}$ were discovered by E. Mathieu [104] in 1861 and M_n is 5-transitive if n is 12 or 24, 4-transitive if n is 11 or 23, and M_{22} and $\operatorname{Aut} M_{22}$ are 3-transitive.

(c) If V is a finite vector space, then the group $AGL(V)$ of affine transformations $\phi_{A,b}$ of V (where $\phi_{A,b} \colon x \mapsto Ax + b$ for $x \in V$, with $b \in V$ and A a nonsingular linear transformation of V) is a 2-transitive permutation group on V. If V is a vector space over the field of order 2 then $AGL(V)$ is 3-transitive on V.

There are several other infinite families of finite 2-transitive permutation groups and a few sporadic examples (see for example [22]). Moreover an old theorem due to Burnside, see [15] p. 202, shows that a finite 2-transitive group G is either a subgroup of an affine group or an *almost simple group*, that is $T \leqslant G \leqslant \operatorname{Aut} T$ for some nonabelian simple group T.

THEOREM 1.2 (BURNSIDE). *If G is a finite 2-transitive permutation group then either*

(a) $G \leqslant AGL(V)$ *for some finite vector space V, or*

(b) $T \leqslant G \leqslant \operatorname{Aut} T$ *for some finite nonabelian simple group T.*

Thus the problem of classifying finite 2-transitive permutation groups splits into two cases, the affine and the almost simple cases. The solution for each case involved the use of the finite simple group classification. The classification of affine 2-transitive groups was completed for soluble groups by Huppert [61] and for insoluble groups by Hering [52], [53], see also Liebeck [90], while the almost simple case was completed by a number of people (see [22]; [103] for $T = A_n$, [37], for T a group of Lie type). A list of the almost simple 2-transitive groups can be found in [22] and of the affine 2-transitive groups in [90]. In particular the only finite 4-transitive groups are the symmetric, alternating, and Mathieu groups.

2. LOW RANK CLASSIFICATIONS

If $G \leqslant \operatorname{Sym}(\Omega)$ then G has a natural faithful action on $\Omega \times \Omega$ given by

$$(\alpha, \beta)^g := (\alpha^g, \beta^g)$$

for $\alpha, \beta \in \Omega$, $g \in G$. Thus $G \leqslant \operatorname{Sym}(\Omega \times \Omega)$. If G is transitive on Ω then one of its orbits in $\Omega \times \Omega$ is the 'diagonal' $\Delta_0 = \{(\alpha, \alpha) \mid \alpha \in \Omega\}$, and, provided $|\Omega| > 1$, G has some other orbits in $\Omega \times \Omega$. Let $\Delta_0, \Delta_1, \ldots, \Delta_{r-1}$ be the G-orbits in $\Omega \times \Omega$, where G is transitive on Ω of degree $|\Omega| > 1$. These sets Δ_i are sometimes called G-*orbitals* and the integer r is called the permutation *rank* of G on Ω. Clearly G has rank 2 on Ω if and only if G is 2-transitive on Ω. Thus the classification of 2-transitive groups G is the classification of permutation groups of lowest possible rank.

Finite permutation groups of rank 3 have received much attention in the literature for several reasons: several of the sporadic simple groups, namely the Higman-Sims, McLaughlan, Suzuki, and Rudvalis simple groups, were discovered as finite rank 3 permutation groups. Also a rank 3 permutation group of even order is a group of automorphisms of a strongly regular graph. Primitive rank 3 permutation groups were investigated extensively by D. G. Higman [55], [56], [57]. Recently, with the aid of the finite simple group classification, the finite rank 3 permutation groups have been completely classified (see Kantor and Liebler [77], Seitz [121], Liebeck [90], and Liebeck and Saxl [97]).

The classification of finite primitive groups of rank at most 5 has been completed except for some sporadic examples and the subgroups of affine groups (by Bannai [8] for such representations of the alternating and symmetric groups, and Cuypers [38]). Finally there are asymptotic results available for finite primitive groups of bounded rank: for a positive integer r, all but finitely many finite primitive permutation groups of rank at most r are known (see [68] for classical groups, [22], Theorem 5.6, and [76] where the result is announced (in Remark 2 following Proposition 1.3)).

3. THE O'NAN-SCOTT THEOREM

Impressive as the results of the previous section are, perhaps the classification of all finite primitive groups of odd degree by Kantor [72] and Liebeck and Saxl [98] is even more striking. How were such results obtained? What techniques were used? And can we expect equally important and impressive results in the future using similar techniques?

Perhaps the single most effective tool for applying the finite simple group classification to problems about permutation groups is the so-called O'Nan-Scott Theorem. This theorem identifies several 'types' of finite primitive permutation groups and, for each type, gives some information about the abstract structure of the group and/or the nature of its permutation action. The present statement of the theorem is a refinement of analyses of finite primitive permutation groups which were part of the folklore of permutation group theory, for example see [7]. O'Nan and Scott independently wrote down statements of the theorem at the Santa Cruz Conference in 1979. The hope was that this analysis would prove to be useful for solving problems about finite primitive permutation groups: such problems could perhaps be solved by solving them separately for primitive groups of each type. And indeed several basic permutation group problems have been solved in this way. The version of the theorem in [120] is incomplete and other versions are available in [3], [22], [83], [93]. We give a statement here which conveys the ideas of the theorem while hiding some of the technical details contained in fuller versions.

O'NAN-SCOTT THEOREM 3.1. *Let $G \leqslant \mathrm{Sym}(\Omega)$ be a primitive group of degree $|\Omega| = n$. Then one of the following holds.*

    (a) AFFINE TYPE: *$G \leqslant AGL(\Omega)$, acting on a finite vector space Ω of prime power order n.*

    (b) DIAGONAL TYPE: *$T^k < G \leqslant T^k(\mathrm{Out}\, T \times S_k)$ where T is a nonabelian simple group and $k \geqslant 2$; the stabilizer in T^k of a point of Ω is a diagonal subgroup* of T^k (that is a subgroup conjugate in $(\operatorname{Aut} T)^k$ to $D := \{(t, \ldots, t) \mid t \in T\}$) and $n = |T^k : D| = |T|^{k-1}$.

(c) ALMOST SIMPLE TYPE: $T \leqslant G \leqslant \operatorname{Aut} T$ for some nonabelian simple group T.

(d) PRODUCT TYPE: G is a primitive subgroup of a wreath product $H \operatorname{wr} S_k$ in product action on $\Omega = \Delta^k$, where H is a primitive subgroup of $\operatorname{Sym}(\Delta)$ of diagonal or almost simple type.

For primitive groups G of affine, diagonal or product type, the theorem gives some useful information about the action of G on Ω. For G a primitive group of affine type, $G = NH$ where N is the group of translations of the vector space Ω and H is an irreducible subgroup of $GL(\Omega)$. Aschbacher's description of maximal subgroups of $GL(\Omega)$ in [1] has often proved to be an effective tool for working with affine type primitive groups. Similarly, Kovács' blow-up construction and decomposition of primitive subgroups of wreath products in product action [85] has proved to be a powerful technique for working with primitive permutation groups of product type which do not have a regular normal subgroup. More details about the structure of primitive groups of diagonal and product types can be found in [47], [48], [84], [86], [87]. The O'Nan-Scott Theorem can often play the role for problems about primitive permutation groups that Burnside's Theorem played for the problem of classifying multiply transitive groups, that is it can often be used to reduce primitive group problems to problems about almost simple groups and affine groups. For example, groups of diagonal or product type usually have large rank, and their suborbits can be described in terms of information about simple groups or smaller primitive groups, respectively. So the small rank classification problems admitted this kind of reduction fairly naturally. In order to illustrate this reduction procedure better, we outline in the next section the strategies used to obtain a classification of the maximal subgroups of A_n and S_n.

4. MAXIMAL SUBGROUPS OF FINITE SYMMETRIC AND ALTERNATING GROUPS

The problem of classifying the maximal (proper) subgroups of finite symmetric and alternating groups was solved by Liebeck, Saxl and the author in [91]. The nature of this classification for the almost simple maximal subgroups needs elaboration, and this will be given later. In order to outline the strategies used we shall consider the problem of classifying the maximal subgroups G of the symmetric group $\operatorname{Sym}(\Omega)$ on a set Ω of n points. If G is intransitive on Ω then G has an orbit Γ in Ω of length k, where $1 \leqslant k \leqslant \frac{n}{2}$. The largest subgroup of $\operatorname{Sym}(\Omega)$ with Γ as an orbit is $G = \operatorname{Sym}(\Gamma) \times \operatorname{Sym}(\Omega \backslash \Gamma) = S_k \times S_{n-k}$, and it is not difficult to prove that this group is a maximal subgroup of $\operatorname{Sym}(\Omega)$ unless $k = \frac{n}{2}$ in which case it has index 2 in the wreath product $S_k \operatorname{wr} S_2$, the subgroup preserving the partition $\{\Gamma | (\Omega \backslash \Gamma)\}$. This process illustrates our approach in the first part of the classification. We consider subgroups of a certain type, in this case intransitive subgroups, identify subgroups of $\text{Sym}(\Omega)$ which are maximal (by inclusion) among the subgroups of this type, and then decide whether or not these subgroups are indeed maximal subgroups of $\text{Sym}(\Omega)$.

Having completed this process for intransitive subgroups we may assume that G is transitive on Ω. If G is imprimitive on Ω then G preserves a partition P of Ω into say b blocks of size a, where $n = ab$, $a > 1$, $b > 1$. The largest subgroup preserving this partition P is $G = S_a \text{ wr } S_b$ and again we can prove that this group is a maximal subgroup of $\text{Sym}(\Omega)$. Thus we may now assume that G is primitive on Ω and continue the process using the O'Nan-Scott Theorem. For primitive groups G of affine type the maximal ones are the full affine groups $G = AGL(d, p)$, where $n = p^d$ with p a prime and $d \geqslant 1$. Mortimer [105] has shown that these groups are maximal subgroups of $\text{Sym}(\Omega)$ when $d \geqslant 2$, and this is also true when $d = 1$. Similarly for primitive groups of diagonal type the maximal ones are of the form $G = T^k.(\text{Out}\, T \times S_k)$ for some nonabelian simple group T and integer $k \geqslant 2$, where $n = |T|^{k-1}$. It is true, but nontrivial to prove, that G is a maximal subgroup of $\text{Sym}(\Omega)$. In fact our proof here relies on the finite simple group classification. If G is a primitive group of product type, and maximal of this type, then $G = S_a \text{ wr } S_b$ in product action, where $n = a^b$, $a \geqslant 5$, $b \geqslant 2$. In [91] a short proof, using the finite simple group classification, is given that these wreath products in product action are maximal subgroups of $\text{Sym}(\Omega)$. Proofs of this, without using the simple group classification, have been given by Jones and Soomro [66], for most values of a and b, and by P. M. Neumann [66], Appendix. (Aschbacher completed this part of the classification and part of the next stage in [2] independently of [91].)

This leaves the case of almost simple primitive subgroups G of $\text{Sym}(\Omega)$. Suppose that G is primitive on Ω and that $T \leqslant G \leqslant \text{Aut}\, T$ for some nonabelian simple group T. Then, for $\alpha \in \Omega$, the stabilizer G_α of α in G is a maximal subgroup of G such that $G_\alpha T = G$. Conversely if G is an almost simple group, say $T \leqslant G \leqslant \text{Aut}\, T$ for a nonabelian simple group T, and if L is a maximal subgroup of G such that $LT = G$, then G acts faithfully by right multiplication on the set $[G : L]$ of right cosets of L in G as an almost simple primitive permutation group. Thus the problem of classifying the finite almost simple primitive permutation groups is equivalent to the problem of classifying all the core-free maximal subgroups of the almost simple groups. And these problems remain unsolved. Similarly the problem of listing all *maximal* almost simple primitive subgroups of $\text{Sym}(\Omega)$ remains open, but our work in [91] shows that this problem is equivalent to the more general one.

Our techniques for handling the almost simple case are quite different from the other cases. We classify all almost simple primitive subgroups of $\text{Sym}(\Omega)$ which are *not* maximal in $\text{Sym}(\Omega)$. Suppose that G is an almost simple primitive subgroup of $\text{Sym}(\Omega)$, with $T \leqslant G \leqslant \text{Aut}\, T$, where T is a nonabelian simple group, and suppose that $G < H < \mathrm{Sym}(\Omega)$ where H is not the alternating group $\mathrm{Alt}(\Omega)$ on Ω. Then H is primitive on Ω since G is, and we consider the various possible types for H.

If H is of affine, diagonal or product type then G can be determined explicitly (see [91], Table II). So we may assume that all subgroups H of $\mathrm{Sym}(\Omega)$ properly containing G, and not equal to $\mathrm{Alt}(\Omega)$ or $\mathrm{Sym}(\Omega)$, are almost simple. If the only such subgroups have socle equal to T then the normalizer $N_{\mathrm{Sym}(\Omega)}(T)$ of T in $\mathrm{Sym}(\Omega)$ is a maximal subgroup of $\mathrm{Alt}(\Omega)$ or $\mathrm{Sym}(\Omega)$. And this group $N_{\mathrm{Sym}(\Omega)}(T)$ is maximal 'of its type' in $\mathrm{Sym}(\Omega)$. Thus we may assume that some such subgroup H has a simple socle different from T. By replacing G, if necessary, by $N_H(T)$, and then replacing H, if necessary, by a subgroup in which G is maximal, we may assume that G is a maximal subgroup of H. Then we have the following situation: H is an almost simple group with two core-free maximal subgroups, G and H_α (where $\alpha \in \Omega$), such that $H = GH_\alpha$ (since G is transitive on Ω) and $G_\alpha = G \cap H_\alpha$ is maximal in G (since G is primitive on Ω). An expression $H = AB$, where A and B are maximal subgroups of H, is called a *maximal factorization* of H.

To complete the classification, all maximal factorizations $H = AB$ of the almost simple groups H were found (see[54] for H an exceptional group of Lie type, and [95] for the other cases). Then we determined the ones for which one of the factors, say A, was almost simple and for which $A \cap B$ was maximal in A. Finally, setting $G = A$ and taking the action of H by right multiplication on the set $\Omega = [H : B]$ of right cosets of B, we obtained a complete list of all pairs (G, H), with $G < H < \mathrm{Sym}(\Omega)$, G and H almost simple primitive groups on Ω with different socles, and G maximal in H.

The information obtained in [91] about almost simple primitive groups goes beyond that required to determine whether or not a given almost simple primitive subgroup G of $\mathrm{Sym}(\Omega)$ is maximal. It provides essentially complete information about the lattice of subgroups of $\mathrm{Sym}(\Omega)$ containing G. In other words it solves the inclusion problem for almost simple primitive groups.

INCLUSION PROBLEM. *Given a primitive subgroup G of $\mathrm{Sym}(\Omega)$, describe all subgroups H of $\mathrm{Sym}(\Omega)$ containing G.*

This inclusion problem has been solved in [114] for the other types of primitive groups. One of the cases which required especially careful treatment was the case where G was a primitive group of product type. Here the blow-up decomposition of Kovács [85] and its properties were essential for the analysis. Another way to pose the inclusion problem is to ask, for a given primitive group $G < \mathrm{Sym}(\Omega)$, for a description of all primitive subgroups of G. This is the appropriate form of the problem when G has a single nonabelian regular normal subgroup. This case has been considered by Förster and Kovács in [48] where an almost complete solution is given.

The results on maximal subgroups of A_n and S_n, and the maximal factorizations of the almost simple groups, are being used to attack other problems about primitive groups, for example the problems of classifying the finite primitive groups with regular subgroups [96], and the problem of classifying 2-closed primitive groups [94]. Both of these problems have applications in algebraic graph theory.

5. Automorphism Groups of Graphs

There is a close connection between permutation groups and automorphism groups of graphs and the interactions between these two areas over the past 25 years have been very fruitful. To understand this connection recall that, if $G \leqslant \mathrm{Sym}(\Omega)$, then G has a natural faithful action on $\Omega \times \Omega$ defined by $(\alpha, \beta)^g := (\alpha^g, \beta^g)$ for $\alpha, \beta \in \Omega$, $g \in G$. Suppose now that G is transitive on Ω of rank $r \geqslant 2$. Then G has r orbits in $\Omega \times \Omega$, called G-orbitals, say $\Delta_0 = \{ (\alpha, \alpha) \mid \alpha \in \Omega \}$, Δ_1, ..., Δ_{r-1}. Given $\alpha \in \Omega$, there is a 1-1 correspondence between the G-orbitals and the G_α-orbits in Ω, namely the G-orbital Δ corresponds to the G_α-orbit $\Delta(\alpha) := \{ \beta \mid (\alpha, \beta) \in \Delta \}$.

To each orbital Δ_i with $1 \leqslant i \leqslant r - 1$ there corresponds a directed graph, which we shall also denote by Δ_i, with vertex set Ω and edge set Δ_i. Clearly G is admitted as a group of automorphisms of Δ_i acting transitively on vertices and edges of Δ_i, and the set of vertices β for which (α, β) is an edge is $\Delta_i(\alpha)$. There is a natural pairing on the G-orbitals, namely the orbital Δ^* paired with Δ is $\Delta^* := \{ (\beta, \alpha) \mid (\alpha, \beta) \in \Delta \}$. If $\Delta^* = \Delta$ then Δ is said to be *self-paired* and the edges of the directed graph Δ fall into pairs of the form $\{ (\alpha, \beta), (\beta, \alpha) \}$. If each such pair of edges is replaced by the corresponding unordered pair $\{\alpha, \beta\}$, interpreted as an edge of an undirected graph, then we obtain an undirected graph Δ admitting G as a group of automorphisms. Thus the study of transitive permutation groups runs parallel to the study of vertex-transitive and edge-transitive directed and undirected graphs. One problem about primitive groups which had a natural interpretation from the point of view of graph theory was the so-called "Sims' conjecture".

SIMS' CONJECTURE 5.1. *There is a function f on the natural numbers having the property that, whenever G is a finite primitive permutation group on Ω, if G_α, $\alpha \in \Omega$, has an orbit in $\Omega - \{\alpha\}$ of length d, then $|G_\alpha| \leqslant f(d)$.*

If $\Delta(\alpha)$ is a G_α-orbit of length d in $\Omega - \{\alpha\}$, then the directed graph associated with the orbital Δ is such that, for each vertex α, there are exactly d vertices β for which (α, β) is an edge, that is the directed graph Δ has out-valency d. Also there are d vertices γ for which (γ, α) is an edge, so the in-valency is also equal to d. The conjecture asserts that, for a finite vertex-primitive, edge-transitive, directed graph with out-valency d, the number of automorphisms fixing a vertex is bounded above by a function of d. Sims proved this when $d = 3$ and suggested in [123] that the conjecture might be true. Much work was done on the conjecture producing partial results (see for example [82], [126], [134]). In particular Thompson [126] showed that there is a function g on the natural numbers, such that for such a group G, there is a prime $p \leqslant d$ and a normal p-subgroup P of G_α with $|G_\alpha : P| \leqslant g(d)$. Using the finite simple group classification the conjecture was proved by Cameron, Saxl, Seitz and the author in [26].

THEOREM 5.2 [26]. *The Sims' Conjecture is true.*

The arguments in [26] show that a function $f(d)$ of the form $\exp(d^2 o(d))$ can be taken, which is comparable in order to the function g obtained by Thompson and Wielandt. The proof given in [26] proceeds as follows. First a proof is given for almost simple primitive groups, and then this, together with the O'Nan-Scott Theorem, is used to complete the proof for all finite primitive groups. In the case of almost simple primitive groups the following 'compactness' argument is used to allow each of the infinite families of finite simple groups to be considered separately. If the conjecture were false for almost simple primitive groups for some d, then there would be an infinite sequence of almost simple primitive groups, G_1, G_2, ... such that for each i a point stabilizer H_i in G_i has an orbit of length d, and $|H_i| \to \infty$ as $i \to \infty$. By Thompson's Theorem, for each i there is a prime $p_i \leqslant d$ and a normal p_i-subgroup P_i of H_i such that $|H_i : P_i| \leqslant g(d)$. Thus $|P_i| \to \infty$ as $i \to \infty$. Since there is only a finite number of distinct primes $p_i \leqslant d$, there must be a prime $p \leqslant d$ such that $p_i = p$ for an infinite number of integers i. So we may assume that $p_i = p$ for all i. Similarly, as there are only a finite number of sporadic finite simple groups and a finite number of infinite families of finite simple groups, there is an infinite family such that the simple socle of G_i is in this family for an infinite number of integers i. Thus we may assume that the simple socles of all the G_i lie in the same infinite family of finite simple groups. The proof then consists of a detailed analysis for each of the infinite families.

Even though the theories of transitive permutation groups and automorphism groups of vertex-transitive and edge-transitive directed graphs can be considered equivalent, as discussed above, the problems which arise from the different points of view are not necessarily the same. For example the condition that the directed graph be connected is a more natural property than the stronger condition that the group be primitive on vertices. There is a conjecture due to Weiss [130] which is analogous to the Sims' Conjecture which is still unsettled.

CONJECTURE 5.3 (WEISS). *There is a function f on the natural numbers having the property that, whenever Δ is a finite undirected graph of valency d with automorphism group G acting transitively on vertices and directed edges, if, for a vertex α, G_α acts primitively on the set $\Delta(\alpha)$ of vertices joined to α, then $|G_\alpha| \leqslant f(d)$.*

In the rest of this section we shall consider just one more family of problems, all to do with s-arc transitivity. Let Δ be a directed (or undirected) graph. For a positive integer s, an s-arc in Δ is a sequence $(\alpha_0, \ldots, \alpha_s)$ of $s+1$ vertices of Δ such that $\alpha_{i-1} \neq \alpha_{i+1}$ for $1 \leqslant i < s$ and, for $1 \leqslant i \leqslant s$, (α_{i-1}, α_i) (or $\{\alpha_{i-1}, \alpha_i\}$) is an edge of Δ. The graph Δ is said to be s-arc transitive if the automorphism group Aut Δ of Δ is transitive on the set of s-arcs of Δ. There is a connection between 2-arc transitivity and the action induced by (Aut $\Delta)_\alpha$ on $\Delta(\alpha)$ for undirected graphs Δ.

LEMMA 5.4. *Let Δ be a finite undirected graph and let G be a group of automorphisms of Δ which is transitive on the vertices of Δ. Then G is transitive on the set of 2-arcs of Δ if and only if, for a vertex α, G_α is 2-transitive on $\Delta(\alpha)$.*

PROOF. Suppose that G is transitive on 2-arcs and let α be a vertex of Δ. Then for any ordered pairs (β, γ) and (β', γ') of distinct vertices of $\Delta(\alpha)$, the sequences (β, α, γ) and $(\beta', \alpha, \gamma')$ are 2-arcs of Δ. Thus for some $g \in G$, $(\beta, \alpha, \gamma)^g = (\beta', \alpha, \gamma')$, that is $g \in G_\alpha$ and $(\beta, \gamma)^g = (\beta', \gamma')$. Thus G_α is 2-transitive on $\Delta(\alpha)$. The proof of the converse is similarly straightforward and is omitted.

Primitive, but not 2-transitive, permutation groups G with a stabilizer G_α which is 2-transitive on one of its orbits $\Delta(\alpha)$, were first studied by Manning in 1927 (see [133], Theorem 17.7). Manning showed that G_α had to have a second orbit $\Sigma(\alpha)$ of length greater than $v := |\Delta(\alpha)|$ and dividing $v(v-1)$. His results were reinterpreted in terms of directed graphs, and were considerably improved by Cameron [20], [21]. Cameron showed that the orbit $\Sigma(\alpha)$ investigated by Manning was the set of points γ such that for some β, $(\alpha, \beta) \in \Delta$ and $(\gamma, \beta) \in \Delta$, and he showed that $|\Sigma(\alpha)| = v(v-1)/k \geqslant 2v$. Moreover the parameter $k = |\Delta(\alpha) \cap \Delta(\gamma)|$, for $\gamma \in \Sigma(\alpha)$, seemed to be very small in most examples; the largest value occurring in an example was 6, and that occurred in the Higman-Sims graph [58]. In [27], [28] it was shown using the classification of 2-transitive groups (and hence the finite simple group classification) that, in the case where the 2-transitive group $G_\alpha^{\Delta(\alpha)}$ induced by G_α on $\Delta(\alpha)$ has a nonabelian simple socle, and where Δ is self-paired, the parameter k is at most 6. This result in its proper context is a result about 2-arc transitive graphs. The primitivity assumption is replaced by the more natural assumption that the graph is connected, and the result is as follows.

THEOREM 5.5 [27], [28]. *Let Δ be a finite connected undirected graph which is not a complete graph and which has a group G of automorphisms acting transitively on vertices such that the stabilizer G_α acts on $\Delta(\alpha)$ as a 2-transitive group with nonabelian simple socle. Then either*

(a) *a pair of vertices at distance 2 is joined by at most 6 paths of length 2, or*

(b) *Δ is a dual orthogonal graph, or*

(c) *Δ is the incidence graph of a known self-dual symmetric 2-design.*

The same problem for the case in which $G_\alpha^{\Delta(\alpha)}$ has an abelian regular normal subgroup has not been solved.

PROBLEM 5.6. *Classify all finite, connected, undirected graphs Δ such that*

(a) *there is a vertex-transitive group G of automorphisms in which the stabilizer G_α acts on $\Delta(\alpha)$ as a 2-transitive group with an abelian regular normal subgroup, and*

(b) *two vertices at distance 2 have more than 6 paths of length 2 joining them.*

Primitive groups G with 2-transitive suborbits were considered in [112] from the point of view of the O'Nan-Scott Theorem. Their classification reduces to the classification of the almost simple, affine, and twisted wreath examples.

PROBLEM 5.7. *Classify all finite primitive permutation groups G such that the stabilizer G_α of a point α is 2-transitive on one of its orbits.*

Another direction taken by the investigations into s-arc transitivity for graphs and directed graphs is the study of such graphs with s large. Interest in s-arc transitive graphs with large s began with work of Tutte [128], [129] who showed that if Δ is a finite s-arc transitive graph of valency 3 then the value of s is at most 5. The problem of bounding s for s-arc transitive graphs of larger valency received a lot of attention, see [49], [130], and was eventually solved by Weiss [131] using the classification of 2-transitive groups.

THEOREM 5.8 [131]. *Let Δ be a finite undirected s-arc transitive graph of valency at least 3. Then $s \leqslant 7$.*

Many constructions of 4-arc transitive and 5-arc transitive graphs of valency 3 are known; these use ideas from combinatorial group theory and several families were discovered after extensive experimentation by computer, see for example [9], [10], [32], [33], [34], [42], [50]. The situation for finite s-arc transitive directed graphs is quite different: for each $s \geqslant 1$ and for each $v \geqslant 3$ there is an infinite number of (pairwise nonisomorphic) finite directed graphs of valency v which are s-arc transitive but not $(s + 1)$-arc transitive, see [115]. Each of these examples has the property that its automorphism group acts imprimitively on vertices. If there is a vertex-primitive 2-arc transitive directed graph then almost certainly there must be such an example with an almost simple automorphism group, see [115], section 4. No such examples have yet been found despite considerable effort by several people.

QUESTION 5.9. *Is there a finite 2-arc transitive directed graph such that the automorphism group is primitive on vertices and is an almost simple group?*

The study of s-arc transitive graphs has also impinged on the study of finite distance transitive graphs. This is another area where the finite simple group classification has been used effectively to bring us close to a complete classification, see [11], [12], [13], [43], [62], [118], [132].

6. AUTOMORPHISM GROUPS OF DESIGNS

A t-(v, k, λ) *design* $\mathcal{D} = (X, \mathcal{B})$ consists of a set X of v points and a set $\mathcal{B}$ of k-element subsets of points called blocks such that each t-element subset of points is contained in exactly λ blocks. The automorphism group G of $\mathcal{D}$ is the subgroup of $\mathrm{Sym}(X)$ which fixes $\mathcal{B}$ setwise. Thus G acts on both X and $\mathcal{B}$, and according to a result of Block (see [60], section 1.6):

LEMMA 6.1. *The number of G-orbits on blocks is greater than or equal to the number of G-orbits on points.*

In particular if G is transitive on blocks then G is transitive on points. The group G can be multiply transitive on points. For example if X is a finite vector space and $\mathcal{B}$ is the set of i-dimensional subspaces of X, for a fixed i, then G contains the full affine group on X. In the literature 2-$(v, k, 1)$ designs with automorphism groups which are 2-transitive on points have been called *2-transitive designs* and these have been classified by Kantor [69] using the simple group classification.

THEOREM 6.2 [69]. *All 2-transitive designs are known.*

A more ambitious program is underway (by Buekenhout and others [14]) to classify all 2-$(v, k, 1)$ designs for which the group G is transitive on flags, that is on incident point-block pairs. Also Delandtsheer [39] has shown that a group G of automorphisms of a 2-$(v, k, 1)$ design, which is primitive on blocks, is an almost simple group unless the design is a projective plane. In the latter case, either the projective plane is Desarguesian, or v is a prime and G is a regular or Frobenius group of order dividing vk or $v(k-1)$: this was shown by Kantor [72]. Recently a lot of work has been done on block-transitive t-designs ($t \geqslant 2$) with no restriction on the parameter λ. Delandtsheer and Doyen [40] showed that such designs are point-primitive provided that the number of points v is large enough.

THEOREM 6.3 [40]. *If $\mathcal{D}$ is a 2-(v, k, λ) design and a group G of automorphisms acts transitively on blocks and imprimitively on points, then $v \leqslant \left(\binom{k}{2} - 1\right)^2$.*

Note that the bound on v is independent of λ. An almost complete classification of designs for which v achieves the bound has been done by Cameron and the author.

THEOREM 6.4 [29]. *If $\mathcal{D}$ is a $2-(v, k, \lambda)$ design with $v = m^2$ where $m = \binom{k}{2} - 1$, $k > 5$ and $k \neq 8$, and some group G of automorphisms of $\mathcal{D}$ is transitive on blocks and imprimitive on points, then one of the following happens:*

(a) *$T^m \leqslant G \leqslant (\text{Aut } T) \text{wr } S_m$, where T is a simple 2-transitive permutation group of degree m and G projects onto a 2-transitive subgroup of S_m;*

(b) *$T_1 \times T_2 \leqslant G \leqslant \text{Aut}(T_1) \times \text{Aut}(T_2)$, where T_1 and T_2 are simple 2-transitive groups of degree m.*

All the designs arising are such that the parameter λ is very large. The number of designs depends on the 2-transitive subgroups of S_m, but if the only 2-transitive groups of degree m are A_m and S_m (and this is the case for 'almost all' m by [25]) then there are exactly 3 non-isomorphic designs satisfying the hypotheses of the theorem. The classification above needs to be completed for $k = 3, 4, 5$ and 8, which are the values of k for which m is a prime power. This has been done in the case $\lambda = 1$.

THEOREM 6.5 [109], [107]. *If $\mathcal{D}$ is a $2-(v, k, 1)$ design with $v = m^2$, where $m = \binom{k}{2} - 1$, and if some group G of automorphisms of $\mathcal{D}$ is transitive on blocks and imprimitive on points, then $\mathcal{D}$ is a $2-(729, 8, 1)$ design, and there are, up to isomorphism, more than 400 such designs.*

The designs arising in the theorem were classified with the help of a computer search, see [107]. Another problem about block transitive t-designs is the existence of examples for large values of t. It was shown by Tierlinck [127] that nontrivial t-designs exist for all t, but the condition that the design be block-transitive puts strong restrictions on t.

THEOREM 6.6 [30]. *There are no nontrivial block-transitive t-designs with $t \geqslant 8$.*

CONJECTURE 6.7 [30]. *There are no nontrivial block-transitive 6-designs.*

7. AUTOMORPHISM GROUPS OF FIELDS

Several problems about Galois groups of algebraic number fields reduce to questions about conjugacy classes in finite groups and thence to questions about finite permutation groups. If H is a subgroup of a finite group G then, in this context, we are often interested in the set

$$\mathcal{C}_G(H) := \cup_{g \in G} H^g$$

which is the union of the G-conjugacy classes of elements of H, and in the set $\mathcal{P}_G(H)$, the union of those G-conjugacy classes in $\mathcal{C}_G(H)$ which consist of elements of prime power order.

One problem concerns relative Brauer groups of global fields (that is algebraic number fields or algebraic function fields in one variable over a finite field). If L is an extension of a field K then the relative Brauer group $B(L/K)$ consists of all Brauer classes of finite-dimensional central simple K-algebras split by L. If K is a finitely generated extension of a global field k and L is a nontrivial finite extension of K, then $B(L/K)$ is infinite. This was proved by Fein, Kantor and Schacher in [44], [46]; see also [45]. Let $\tilde{L}$ be the Galois hull of L/K, and let G be the Galois group of $\tilde{L}/K$ and H the fixed group of L. They show that if $B(L/K)$ is finite then $\mathcal{P}_G(H) = G$. Now the equality $\mathcal{P}_G(H) = G$, for a finite group G with subgroup H, is equivalent to the property that the transitive permutation group induced by G by right multiplication on the set $[G : H]$ of right cosets of H in G contains no fixed-point-free elements of prime power order. It is shown in [44] that no finite transitive permutation group of degree greater than 1 has this property.

THEOREM 7.1 [44]. *If G is a transitive permutation group on a finite set Ω of degree $|\Omega| > 1$ then for some prime p there is a p-element in G which fixes no point of Ω.*

In [44], Theorem 7.1 is proved by first reducing to the case where G is primitive on Ω, then, using the O'Nan-Scott Theorem, reducing to the case where G is simple, and then the finite simple group classification is used to complete the proof. No other proof has as yet been found. As we discussed above, Theorem 7.1 is equivalent to the assertion that $\mathcal{P}_G(H) \neq \mathcal{P}_G(G)$ whenever H is a proper subgroup of G. Guralnick [51] considered a generalization of this situation, namely the case of two subgroups K, H of a finite group G with $\mathcal{P}_G(H) \supseteq \mathcal{P}_G(K)$ (so G is replaced by a subgroup K). He dealt with the case where G induces an alternating or symmetric group in its action on the cosets of K.

THEOREM 7.2 [51]. *Suppose that K is a subgroup of a group G of index n and that $G/\mathrm{core}_G(H) \simeq A_n$, $n > 5$, or S_n, where $\mathrm{core}_G(H) = \cap_{g \in G} H^g$. If $H \leqslant G$ satisfies $\mathcal{P}_G(H) \supseteq \mathcal{P}_G(K)$ then $H \geqslant K^g$ for some $g \in G$.*

When $n = 1$ this is just Theorem 7.1, and when $n = 2$ it is a result of Saxl [119]. One crucial result underlying this theorem is the following result about simple groups. Its proof requires the simple group classification, and is essentially the proof given in [119] for the analogous result with $\mathcal{C}_G(\ )$ instead of $\mathcal{P}_G(\ )$.

THEOREM 7.3 ([51], Theorem 2.1 and [119]). *Let S be a finite nonabelian simple group. If $H \leqslant S \leqslant G \leqslant \mathrm{Aut}\, S$ and $H \neq S$ then $\mathcal{P}_G(H) \neq \mathcal{P}_G(S)$.*

Theorem 7.2 has applications to relative Brauer groups as discussed above, and also to Kronecker classes of fields which we shall discuss now. For an algebraic number field k and a finite extension K of k, the *Kronecker set* $D(K/k)$ of K over k is the set of all primes of k which have a degree one factor in K. Two extensions K_1 and K_2 of k are said to be *Kronecker equivalent* relative to k if their Kronecker sets over k have finite symmetric difference. If M/k is a Galois extension and K_1, K_2 are subfields of M then the Kronecker equivalence of K_1 and K_2 is expressible in terms of Galois groups: namely if G is the Galois group of M/k and U_1, U_2 are the fixed groups of K_1 and K_2 respectively, then K_1 and K_2 are Kronecker equivalent relative to k if and only if $C_G(U_1) = C_G(U_2)$, see [64]. Jehne made an extensive investigation of Kronecker equivalence in [64]. If K_1/k is a Galois extension of degree n then U_1 is normal in G of index n, and if K_2 is Kronecker equivalent to K_1 relative to k, $K_2 \neq K_1$, then $K_2 \supset K_1$, $U_2 < U_1$, and $C_G(U_2) = U_1$. Then in the permutation representation induced by G on the right cosets of U_2, the subgroup U_1 has n orbits which are permuted regularly by G, and the condition $C_G(U_2) = U_1$ is equivalent to the condition that U_1 contains no fixed-point-free elements. If in addition K_2 is an atomic extension of K_1, that is if U_2 is a maximal subgroup of U_1, then U_1 is primitive on each of its n orbits.

Jehne and Klingen [63], [64], [79], [80] were particularly interested in quadratic extensions K_1/k; they believed that K_1 would be Kronecker equivalent only to itself. They reduced the problem to a consideration of groups G with a nonabelian simple normal subgroup U_1 of index 2, and a maximal subgroup U_2 such that $C_G(U_2) = U_1$. Saxl [119] showed (and it follows from Theorem 7.2) that the only possibility is $U_2 = U_1$. This quadratic case is exceptional in that there are many examples of groups G with a normal subgroup U_1 of index $n > 2$ and with $U_2 < U_1$ satisfying $C_G(U_2) = U_1$. In particular there are nontrivial examples for each n with $3 \leqslant n \leqslant 8$. Using Theorem 7.3 we can classify explicitly all possibilities for $G/\operatorname{core}_G(U_2)$ when $n \leqslant 8$.

THEOREM 7.4 [113], [116], [117]. *Let G be a finite group with normal subgroup U_1 of index n, where $3 \leqslant n \leqslant 8$. Suppose that U_2 is a proper subgroup of U_1 such that $C_G(U_2) = U_1$ and $\operatorname{core}_G(U_2) = 1$. Then there are only a finite number of possibilities for G, U_1, U_2, and these are known explicitly.*

Klingen [81] has shown, in the case $n = 4$, that each of the two possible groups G can be realized as Galois groups of field extensions giving a quartic Galois extension K_1/k Kronecker equivalent to a cubic extension K_2 of K_1. Further he has shown that the problem of finding a proper extension K_2 of a given quartic Galois extension K_1/k, with K_1 and K_2 Kronecker equivalent, depends on the nature of the extension K_1/k. He has found necessary and sufficient number theoretic conditions for such an extension to be possible. So the problem of constructing Kronecker equivalent extensions depends heavily on number theory as well as on group theory.

There is a basic open problem in this area posed by P. M. Neumann.

PROBLEM 7.5. *Let G be a finite group with a normal subgroup U_1 of index n, and let U_2 be a subgroup of U_1 such that $C_G(U_2) = U_1$. Prove that the index $|U_1 : U_2|$ is bounded above by a function of n.*

This problem has been solved by Neumann and the author in the case where U_2 is a maximal subgroup of U_1. A (positive) solution of the problem would imply that, for an extension K_1/k of degree m, all extensions K_2/k Kronecker equivalent to K_1 must have degree bounded above by a function of m.

8. PERMUTATION GROUP ALGORITHMS

Sims [122], [124], [125] introduced the concept of a base and strong generating set for a permutation group G on a finite set Ω. A *base* for G is a sequence $B = (\beta_1, \ldots, \beta_k)$ of points of Ω such that the only element of G fixing each of $\beta_1, \ldots, \beta_k$ is the identity permutation, and a *strong generating set* for G relative to B is a generating set S for G such that for each $1 \leqslant i \leqslant k$ the subset of S fixing $\beta_1, \ldots, \beta_i$ is a generating set for the stabilizer $G_{\beta_1, \ldots, \beta_i}$. Sims developed a practical algorithm for computing a base and strong generating set for a permutation group $G \leqslant \text{Sym}(\Omega)$, the idea for which was based on Schreier's Lemma. (Schreier's Lemma determines a generating set for a subgroup H of a group G, given a generating set for G and a transversal for H in G, see [125] or [89].) Then using this base and strong generating set Sims produced algorithms for finding $|G|$, for testing membership in G of a given permutation of Ω, and for making a random selection of an element of G. He also gave algorithms, using a base and strong generating set, for computing centralizers of elements, conjugacy classes, and intersections of subgroups [125]. It is these algorithms of Sims on which most of the currently used algorithms for computing with finite permutation groups are based. The development of many permutation group algorithms was strongly influenced by Sims' work, see for example [4], [16], [17], [18], [19], [31], [35], [36], [59], [65], [88], [89], and they have been implemented to form part of the group theory system CAYLEY [31]. Several of the basic algorithms have also been implemented as part of the new system GAP [108] in Aachen. The paper [31] of Cannon contains a more detailed survey of the area. Sims' paper [124] contains a list of all primitive permutation groups of degree up to 20; Sims (and others) have extended this list to primitive groups of degree up to 50, and Dixon and Mortimer [41] have given a basic description of the primitive non-affine groups of degree up to 1000. More recently the O'Nan-Scott Theorem and results about primitive groups depending on the finite simple group classification have been used to develop new algorithms for computing with permutation groups. We shall mention just a few. Cameron and Cannon extended the ideas of their probabilistic recognition algorithm [24] for alternating and symmetric groups to an algorithm [23] for recognizing 2-transitive groups. In [70], [71], [74], [78] Kantor and Taylor gave polynomial time algorithms for finding elements of prime order and Sylow subgroups of permutation groups. Luks [101] showed how to compute the composition factors of a permutation group in polynomial time. Neumann [106] developed practical algorithms for finding (i) the socle of a primitive group, (ii) the p-core $O_p(G)$ for an arbitrary permutation group G, and (iii) the composition factors of an arbitrary permutation group. Also he produced a test for simplicity of permutation groups. Then, relying even more heavily on properties of the finite simple groups, Kantor [73], [75] developed some alternative fast algorithms for finding socles, p-cores, and composition factors, and for testing the simplicity of permutation groups. Finally Luks and others [5], [6], [100], [102] have made an extensive investigation of the complexity of various permutation group algorithms under parallel computation.

REFERENCES

[1] Aschbacher, M., 'On the maximal subgroups of the finite classical groups', *Invent. Math.* **76** (1984), 469–514.

[2] Aschbacher, M., 'Maximal subgroups of finite alternating and symmetric groups' (unpublished manuscript, 1985).

[3] Aschbacher, M., and Scott, L., 'Maximal subgroups of finite groups', *J. Algebra* **92** (1985), 44–80.

[4] Atkinson, M. D., 'An algorithm for finding the blocks of a permutation group', *Math. Comp.* **29** (1975), 911–913.

[5] Babai, L., Luks, E. M., and Seress, A., 'Permutation groups in NC', in *Proc. 19th ACM Symposium on Theory of Computing*, to appear.

[6] Babai, L., Luks, E. M., and Seress, A., 'On managing permutation groups in $O(n^4 log^c n)$', *Proc. 28th IEEE FOCS* (1988), 272–282.

[7] Baer, R., 'Classes of finite groups and their properties', *Illinois J. Math.* **1** (1957), 115–187.

[8] Bannai, E., 'Maximal subgroups of low rank of finite symmetric and alternating groups', *J. Fac. Sci. Univ. Tokyo* **18** (1972), 475–486.

[9] Biggs, N., 'Presentations for cubic graphs', in *Computational Group Theory*, Proceedings, Durham, 1982; ed. by Michael D. Atkinson, pp. 57–63 (Academic Press, London, New York, 1984).

[10] Biggs, N., and Hoare, M., 'The sextet construction for cubic graphs', *Combinatorica* **3** (1983), 153–165.

[11] van Bon, J., and Cohen, A. M., 'Prospective classification of distance-transitive graphs', in *Proceedings of the Combinatorics 1988 Conf., Ravello*, pp. 1–9.

[12] van Bon, J., and Cohen, A. M., 'Linear groups and distance-transitive graphs', *European J. Combin.* **10** (1989), 399–412.

[13] Brouwer, A. E., Cohen, A. M., and Neumaier, A., *Distance Regular Graphs* (Springer-Verlag, Berlin, 1989).

[14] Buekenhout, F., Delandtsheer, A., Doyen, J., Kleidman, P. B., Liebeck, M. W., and Saxl, J., 'Linear spaces with flag-transitive automorphism groups', *Geom. Dedicata* (to appear).

[15] Burnside, W., *Theory of Groups of Finite Order* (Cambridge Univ. Press, Cambridge, 1911).

[16] Butler, G., 'Computing in permutation and matrix groups II; backtrack algorithm', *Math. Comp.* **39** (1982), 671–680.

[17] Butler, G., 'Computing normalizers in permutation groups', *J. Algorithms* **4** (1983), 163–175.

[18] Butler, G., and Cannon, J. J., 'Computing in permutation and matrix groups I: normal closure, commutator subgroups, series', *Math. Comp.* **39** (1982), 663–670.

[19] Butler, G., and Cannon, J. J., 'Computing in permutation and matrix groups III: Sylow subgroups', *J. Symbolic Comput.* **8** (1989), 241–252.

[20] Cameron, P. J., 'Permutation groups with multiply transitive suborbits, I', *Proc. London Math. Soc.* (3) **25** (1972), 427–440.

[21] Cameron, P. J., 'Suborbits in transitive permutation groups', in *Combinatorics, Part 3: Combinatorial group theory*, ed. by M. Hall Jr and J. H. van Lint, Math. Centre Tracts **57**, pp. 98–129 (Math. Centrum, Amsterdam, 1974).

[22] Cameron, P. J., 'Finite permutation groups and finite simple groups', *Bull. London Math. Soc.* **13** (1981), 1–22.

[23] Cameron, P. J., and Cannon, J. J., 'Recognizing doubly transitive groups' (preprint, 1985).

[24] Cameron, P. J., and Cannon, J. J., 'Fast recognition of alternating and symmetric groups' (preprint, 1986).

[25] Cameron, P. J., Neumann, P. M., and Teague, D. N., 'On the degrees of primitive permutation groups', *Math. Z.* **180** (1982), 141–149.

[26] Cameron, P. J., Praeger, C. E., Saxl, J., and Seitz, G. M., 'On the Sims conjecture and distance transitive graphs', *Bull. London Math. Soc.* **15** (1983), 499–506.

[27] Cameron, P. J., and Praeger, C. E., 'Graphs and permutation groups with projective subconstituents', *J. London Math. Soc.* (2) **25** (1982), 62–74.

[28] Cameron, P. J., and Praeger, C. E., 'On 2-arc transitive graphs of girth 4', *J. Combin. Theory Ser. B* **35** (1983), 1–11.

[29] Cameron, P. J., and Praeger, C. E., 'Block-transitive designs, I: point-imprimitive designs', *Discrete Math.* (submitted).

[30] Cameron, P. J., and Praeger, C. E., in preparation.

[31] Cannon, J. J., 'A computational toolkit for finite permutation groups', in *Proceedings of the Rutgers Group Theory Year 1983–1984*, ed. by M. Aschbacher et al., pp. 1–18 (Cambridge University Press, New York, 1984).

[32] Conder, M., 'An infinite family of 5-arc transitive cubic graphs', *Ars Combinatoria* **25A** (1988), 95–108.

[33] Conder, M., 'An infinite family of 4-arc transitive cubic graphs each with girth 12', *Bull. London Math. Soc.* **21** (1989), 375–380.

[34] Conder, M., and Lorimer, P., 'Automorphism groups of symmetric graphs of valency 3', *J. Combin. Theory Ser. B* **47** (1989), 60–72.

[35] Cooperman, G., and Finkelstein, L. A., 'A strong generating test and short presentations for permutation groups', *J. Symbolic Comput.* (to appear).

[36] Cooperman, G., Finkelstein, L., and Purdom, P. W., 'Fast group membership using a strong generating test for permutation groups', *Proc. Computers and Math.* (to appear).

[37] Curtis, C. W., Kantor, W. M., and Seitz, G., 'The 2-transitive permutation representations of the finite Chevalley groups', *Trans. Amer. Math. Soc.* **218** (1976), 1–57.

[38] Cuypers, H., *Geometries and permutation groups of small rank* (Rijksuniversiteit, Utrecht, 1989).

[39] Delandtsheer, A., 'Line-primitive automorphism groups of finite linear spaces', *European J. Combin.* **10** (1989), 161–169.

[40] Delandtsheer, A., and Doyen, J., 'Most block transitive *t*-designs are point primitive', *Geom. Dedicata* **29** (1989), 397–410.

[41] Dixon, J. D., and Mortimer, B., 'The primitive permutation groups of degree less than 1000', *Math. Proc. Cam. Phil. Soc.* **103** (1988), 213–238.

[42] Djoković, D. Ž., and Miller, G. L., 'Regular groups of automorphisms of cubic graphs', *J. Combin. Theory Ser. B* **29** (1980), 195–230.

[43] Faradzev, I. A., and Ivanov, A. A., 'Distance-transitive representations of the groups G, $PSL_2(q) \leqslant G \leqslant P\Gamma L_2(q)$' (preprint, 1987).

[44] Fein, B., Kantor, W. M., and Schacher, M., 'Relative Brauer groups, II', *J. Reine Angew. Math.* **328** (1981), 39–57.

[45] Fein, B., and Schacher, M., 'Relative Brauer groups, I', *J. Reine Angew. Math.* **321** (1981), 179–194.

[46] Fein, B., and Schacher, M., 'Relative Brauer groups, III', *J. Reine Angew. Math.* **335** (1982), 37–39.

[47] Förster, P., 'On primitive groups with regular normal subgroups' (preprint, 1985).

[48] Förster, P., and Kovács, L. G., 'On primitive groups with a single nonabelian regular normal subgroup' (preprint, 1989).

[49] Gardiner, A., 'Symmetry conditions in graphs', in *Surveys in Combinatorics*, ed. by B. Bollobás, London Math. Soc. Lecture Note Ser. **38**, pp. 22–43 (Cambridge University Press, Cambridge, 1979).

[50] Goldschmidt, D. M., 'Automorphisms of trivalent graphs', *Ann. of Math.* **111** (1980), 377–406.

[51] Guralnick, R. M., 'Zeros of permutation characters with applications to prime splitting and Brauer groups' (preprint, 1988).

[52] Hering, C., 'Transitive linear groups and linear groups which contain irreducible subgroups of prime order', *Geom. Dedicata* **2** (1974), 425–460.

[53] Hering, C., 'Transitive linear groups and linear groups which contain irreducible subgroups of prime order, II', *J. Algebra* **93** (1985), 151–164.

[54] Hering, C., Liebeck, M. W., and Saxl, J., 'The factorizations of the finite exceptional groups of Lie type', *J. Algebra* **106** (1987), 517–527.

[55] Higman, D. G., 'Finite permutation groups of rank 3', *Math. Z.* **86** (1964), 145–156.

[56] Higman, D. G., 'Primitive rank 3 groups with a prime subdegree', *Math. Z.* **91** (1966), 70–86.

[57] Higman, D. G., 'Intersection matrices for finite permutation groups', *J. Algebra* **6** (1967), 22–42.

[58] Higman, D. G., and Sims, C. C., 'A simple group of order 44,352,000', *Math. Z.* **105** (1968), 110–113.

[59] Holt, D. F., 'The computation of normalizers in permutation groups', *J. Symbolic Comput.* (to appear).

[60] Hughes, D. R., and Piper, F. C., *Design Theory* (Cambridge University Press, Cambridge, 1985).

[61] Huppert, B., 'Zweifach transitive, auflösbare Permutationsgruppen', *Math. Z.* **68** (1957), 126–150.

[62] Ivanov, A. A., 'Distance-transitive graphs and their classification' (preprint, 1989).

[63] Jehne, W., 'Kronecker classes of atomic extensions', *Proc. London Math. Soc.* (3) **34** (1977), 32–64.

[64] Jehne, W., 'Kronecker classes of algebraic number fields', *J. Number Theory* **9** (1977), 279–320.

[65] Jerrum, M., 'A compact representation for permutation groups', *J. Algorithms* **7** (1986), 60–78.

[66] Jones, G. A., and Soomro, K. D., 'The maximality of certain wreath products in alternating and symmetric groups', *Quart. J. Math. Oxford* (2) **37** (1986), 419–435.

[67] Kantor, W. M., 'Permutation representations of the finite classical groups of small degree or rank', *J. Algebra* **60** (1979), 158–168.

[68] Kantor, W. M., 'Some consequences of the classification of finite simple groups', in *Finite groups—coming of age*, ed. by John McKay, Contemporary Math. **45**, pp. 159–173 (Amer. Math. Soc., Providence, 1985).

[69] Kantor, W. M., 'Homogeneous designs and geometric lattices', *J. Combin. Theory Ser. A* **38** (1985), 66–74.

[70] Kantor, W. M., 'Polynomial-time algorithms for finding elements of prime order and Sylow subgroups', *J. Algorithms* **6** (1985), 478–514.

[71] Kantor, W. M., 'Sylow's theorem in polynomial time', *J. Comput. System Sci.* **30** (1985), 359–394.

[72] Kantor, W. M., 'Primitive permutation groups of odd degree, and an application to finite projective planes', *J. Algebra* **106** (1987), 15–45.

[73] Kantor, W. M., 'Algorithms for computing in permutation groups' (preprint, 1988).

[74] Kantor, W. M., 'Algorithms for Sylow p-subgroups and solvable groups', in *Computers in Algebra*, pp. 77–90 (Marcel Dekker, New York, 1988).

[75] Kantor, W. M., 'Finding composition factors of permutation groups of degree $n \leqslant 10^6$', *J. Symbolic Comput.* (to appear).

[76] Kantor, W. M., Liebeck, M. W., and Macpherson, H. D., '$\aleph_0$-categorical structures smoothly approximated by finite substructures' (preprint, 1988).

[77] Kantor, W. M., and Liebler, R. A., 'The rank 3 permutation representations of the finite classical groups', *Trans. Amer. Math. Soc.* **71** (1982), 1–71.

[78] Kantor, W. M., and Taylor, D. E., 'Polynomial-time versions of Sylow's theorem', *J. Algorithms* (to appear).

[79] Klingen, N., 'Zahlkörper mit gleicher Primzerlegung', *J. Reine Angew. Math.* **299/300** (1978), 342–384.

[80] Klingen, N., 'Atomare Kronecker-Klassen mit speziellen Galoisgruppen', *Abhandl. Math. Sem. Hamburg* **48** (1979), 42-53.

[81] Klingen, N., 'Rigidity of decomposition laws and number fields', *J. Austral. Math. Soc. Ser. A* (to appear).

[82] Knapp, W., *Über einige Fragen aus der Theorie der endlichen Permutationsgruppen, die sich in Zusammenhang mit einer Vermutung von Sims stellen* (Habilitationsschrift, Universität Tübingen, 1977).

[83] Kovács, L. G., 'Maximal subgroups in composite finite groups', *J. Algebra* **99** (1986), 114-131.

[84] Kovács, L. G., 'Primitive permutation groups of simple diagonal type', *Israel J. Math.* **63** (1988), 119-127.

[85] Kovács, L. G., 'Primitive subgroups of wreath products in product action', *Proc. London Math. Soc.* (3) **58** (1989), 306–322.

[86] Kovács, L. G., 'Twisted wreath products as primitive permutation groups' (in preparation).

[87] Lafuente, J., 'Grupos primitivos con subgrupos maximales penquenos', *Publ. Sec. Mat. Univ. Auton. Barcelona* **29** (1985), 154-161.

[88] Leedham-Green, C. R., Praeger, C. E., and Soicher, L. H., 'Algorithms for finding the kernel of a group homomorphism', *J. Symbolic Comput.* (to appear).

[89] Leon, J. S., 'On an algorithm for finding a base and strong generating set for a group given by generating permutations', *Math. Comp.* **35** (1980), 941–974.

[90] Liebeck, M. W., 'The affine permutation groups of rank 3', *Proc. London Math. Soc.* (3) **54** (1987), 477–516.

[91] Liebeck, M. W., Praeger, C. E., and Saxl, J., 'A classification of the maximal subgroups of the finite alternating and symmetric groups', *J. Algebra* **111** (1987), 365-383.

[92] Liebeck, M. W., Praeger, C. E., and Saxl, J., 'On maximal subgroups and maximal factorizations of almost simple groups', *Proc. Symp. Pure Math.* **47** (1987), 449-454.

[93] Liebeck, M. W., Praeger, C. E., and Saxl, J., 'On the O'Nan-Scott Theorem for finite primitive permutation groups', *J. Austral. Math. Soc. Ser. A* **44** (1988), 389–396.

[94] Liebeck, M. W., Praeger, C. E., and Saxl, J., 'On the 2-closures of primitive permutation groups', *J. London Math. Soc.* (2) **37** (1988), 241-252.

[95] Liebeck, M. W., Praeger, C. E., and Saxl, J., 'The factorizations of the finite simple groups and their automorphism groups', *Memoirs Amer. Math. Soc.* **432** (1990) (to appear).

[96] Liebeck, M. W., Praeger, C. E., and Saxl, J., 'Finite primitive permutation groups containing regular subgroups' (in preparation).

[97] Liebeck, M. W., and Saxl, J., 'Some recent results on finite permutation groups', in *Proceedings of the Rutgers Group Theory Year 1983–1984*, ed. by M. Aschbacher et al., pp. 53–61 (Cambridge University Press, New York, 1984).

[98] Liebeck, M. W., and Saxl, J., 'The primitive permutation groups of odd degree', *J. London Math. Soc.* (2) **31** (1985), 250–264.

[99] Liebeck, M. W., and Saxl, J., 'The finite primitive permutation groups of rank three', *Bull. London Math. Soc.* **18** (1986), 165–172.

[100] Luks, E. M., 'Parallel algorithms for permutation groups and graph isomorphism', *Proc. 27th IEEE FOCS* (1986), 292–302.

[101] Luks, E. M., 'Computing the composition factors of a permutation group in polynomial time', *Combinatorica* **7** (1987), 87–99.

[102] Luks, E. M., and McKenzie, P., 'Fast parallel computation with permutation groups', *Proc. 26th IEEE FOCS* (1985), 505–514.

[103] Maillet, E., 'Sur les isomorphes holoédriques et transitifs des groupes symetriques ou alternés', *J. Math. Pures Appl.* (5) **1** (1895), 5–34.

[104] Mathieu, E., 'Mémoire sur l'étude des fonctions de plusieurs quantités', *J. Math. Pures Appl.* (2) **6** (1861), 241–323.

[105] Mortimer, B., 'Permutation groups containing affine groups of the same degree', *J. London Math. Soc.* **15** (1977), 445–455.

[106] Neumann, P. M., 'Some algorithms for computing with finite permutation groups', in *Groups— St Andrews 1985*, ed. by C. M. Campbell and E. F. Robertson, London Math. Soc. Lecture Note Ser. 121, pp. 59–92 (Cambridge University Press, Cambridge, 1986).

[107] Nickel, W., Niemeyer–Nickel, A. C., O'Keefe, C., Penttila, T., and Praeger, C. E., 'The block-transitive, point-imprimitive 2-(729,8,1) designs' (preprint, 1990).

[108] Nickel, W., Niemeyer, A., and Schönert, M., GAP, *getting started and reference manual* (RWTH Aachen, 1988).

[109] O'Keefe, C., Penttila, T., and Praeger, C. E., 'On block-transitive, point-imprimitive designs', *Discrete Math.* (submitted).

[110] Praeger, C. E., 'Finite simple groups and finite primitive permutation groups', *Bull. Austral. Math. Soc.* **28** (1983), 355–366.

[111] Praeger, C. E., 'Symmetric graphs and the classification of the finite simple groups', in *Groups— Korea 1983*, ed. by A. C. Kim and B. H. Neumann, Lecture Notes in Math. **1098**, pp. 99–110 (Springer-Verlag, Berlin, 1984).

[112] Praeger, C. E., 'Primitive permutation groups with doubly transitive subconstituents', *J. Austral. Math. Soc. Ser. A* **45** (1988), 66–77.

[113] Praeger, C. E., 'Covering subgroups of groups and Kronecker classes of fields', *J. Algebra* **118** (1988), 455–463.

[114] Praeger, C. E., 'The inclusion problem for finite primitive permutation groups', *Proc. London Math. Soc.* (3) **60** (1990), 69–88.

[115] Praeger, C. E., 'Highly arc-transitive digraphs', *European J. Combin.* **10** (1989), 281–292.

[116] Praeger, C. E., 'Kronecker classes of field extensions of small degree', *J. Austral. Math. Soc. Ser. A* (to appear).

[117] Praeger, C. E., 'On octic extensions and a problem in group theory', in *Group Theory*, Proceedings of the Singapore Group Theory Conference held at the National University of Singapore, 1987; ed. by Kai Nah CHENG and Yu Kiang LEONG, pp. 443–463 (de Gruyter, Berlin, New York, 1989).

[118] Praeger, C. E., Saxl, J., and Yokoyama, K., 'Distance transitive graphs and finite simple groups', *Proc. London Math. Soc.* (3) **55** (1987), 1–21.

[119] Saxl, J., 'On a question of W. Jehne concerning covering subgroups of groups and Kronecker classes of fields', *J. London Math. Soc.* (2) **38** (1988), 243–249.

[120] Scott, L. L., 'Representations in characteristic p', in *The Santa Cruz conference on finite groups*, ed. by Bruce Cooperstein and Geoffrey Mason, Proc. Symposia in Pure Math. **37**, pp. 318–331 (Amer. Math. Soc., Providence, 1980).

[121] Seitz, G. M., 'Small rank permutation representations of finite Chevalley groups', *J. Algebra* **28** (1974), 508–517.

[122] Sims, C. C., 'Determining the conjugacy classes of a permutation group', in *Computers in Algebra and Number Theory*, SIAM-AMS Proc. **4**, pp. 191–195, 1970.

[123] Sims, C. C., 'Graphs and finite permutation groups', *Math. Z.* **95** (1967), 76–86.

[124] Sims, C. C., 'Computational methods in the study of permutation groups', in *Computational Problems in Abstract Algebra*, ed. by J. Leech, pp. 169–183 (Pergamon Press, New York, 1970).

[125] Sims, C. C., 'Computation with permutation groups', in *Proc. Second Symp. Symbolic and Algebraic Manipulation* (Assn. Computing Mach., New York, 1971).

[126] Thompson, J. G., 'Bounds for the orders of maximal subgroups', *J. Algebra* **14** (1970), 135–138.

[127] Tierlinck, L., 'Non-trivial t-designs without repeated blocks exist for all t', *Discrete Math.* **65** (1987), 301–311.

[128] Tutte, W. T., 'A family of cubical graphs', *Proc. Cambridge Philos. Soc.* **43** (1947), 459–474.

[129] Tutte, W. T., 'On the symmetry of cubic graphs', *Canad. J. Math.* **11** (1959), 621–624.

[130] Weiss, R., 's-transitive graphs', in *Algebraic Methods in Graph Theory*, Coll. Math. Soc. János Bolyai **25**, pp. 827–847, 1984.

[131] Weiss, R., 'The non-existence of 8-arc transitive graphs', *Combinatorica* **1** (1981), 309–311.

[132] Weiss, R., 'Distance-transitive graphs and generalised polygons', *Arch. Math. (Basel)* **45** (1985), 186–192.

[133] Wielandt, H., *Finite Permutation Groups* (Academic Press, New York, 1964).

[134] Wielandt, H., *Subnormal subgroups and permutation groups* (Ohio State University, 1971).

Department of Mathematics
University of Western Australia
NEDLANDS WA 6009
Australia

RESIDUALLY FINITE GROUPS

Dan Segal

There are many well known theorems which assert that one or another kind of infinite group is residually finite. Only recently, however, have results begun to emerge which take residual finiteness as an (explicit or implicit) hypothesis. Some striking conclusions have been reached, and some challenging new problems raised. This work will be surveyed under three headings: § 1 Growth of finite-index subgroups; § 2 Zeta functions and Poincaré series; § 3 Uniform finiteness conditions. Underlying many of the results is the theory of p-adic analytic groups; some aspects of this are sketched in § 4.

An excellent survey of results and methods under 1 and 2 has recently been given by Avinoam Mann [M]: the reader is encouraged to peruse this for further details and another point of view.

1. GROWTH OF FINITE-INDEX SUBGROUPS

To be residually finite, for an infinite group, means to have a lot of subgroups of finite index. Now it is clear that some such groups, for example, the infinite cyclic group, have 'only just enough' subgroups of finite index, while others, such as the non-abelian free groups, have an abundance of them. We can make this distinction precise by defining the *growth rate* of finite-index subgroups of a group G as follows:

$$\alpha(G) = \limsup_{n \to \infty} \frac{\log s_n(G)}{\log n}$$

where $s_n(G)$ denotes the number of subgroups of index at most n in G. It is easy to see that

$$\alpha(G) = \inf \left\{ \alpha \mid (\exists c) \, (\forall n) \; s_n(G) \leqslant c n^\alpha \right\},$$

so $\alpha(G)$ is in effect the *degree of polynomial growth* of $s_n(G)$, if indeed this sequence grows at most polynomially with n, while $\alpha(G) = \infty$ if $s_n(G)$ grows faster than any polynomial in n. Of course, this is only interesting if we know that $s_n(G)$ is finite for all n: this is certainly so if G is finitely generated (it is also true if G has finite 'upper rank', defined below; see [MS], Lemma 3.5).

The group G has PSG ('polynomial subgroup growth') if $\alpha(G)$ is finite. I observed in [S1] that *every soluble group of finite rank has PSG*, and raised

PROBLEM 1. Characterise the finitely generated residually finite groups which have *PSG*.

A special case of this problem was also solved in [**S1**]: a *finitely generated residually nilpotent soluble group has PSG if and only if it is a soluble minimax group*. (It is a well-known result, due to Derek Robinson, [**R**], that the finitely generated soluble groups of finite rank are just the finitely generated soluble minimax groups.) Introducing powerful new methods, Lubotzky and Mann [**LM3**] prove the same theorem *without assuming solubility*. These methods seem likely to have far-reaching applications, and I would like to indicate briefly the main ingredients.

Step 1: If G is finitely generated and has *PSG* then, for each prime p, the pro-p completion $\widehat{G}_p$ of G is p-adic analytic. From this, it is deduced that $\widehat{G}_p$ is a linear group over the p-adic numbers.

Step 2: If G is a finitely generated linear group over a field of characteristic zero and G is not soluble-by-finite, then G has a quotient, with the same property, which is linear over the ring $R = \mathbf{Z}[1/m]$ for some $m \in \mathbf{N}$ (this is an application of results of Jordan, Zassenhaus, Mal'cev and Platonov).

Step 3: An important recent theorem of M. Nori [**N**] and B. Weisfeiler [**MVW**]: if $G \leqslant GL_d(R)$, with R as above, and the Zariski closure $\overline{G}$ of G is a simply connected semi-simple algebraic group, then the *congruence closure* [†] of G in $GL_d(R)$ has finite index in $\overline{G}(R) = \overline{G} \cap GL_d(R)$.

Step 4: An application of the *Prime Number Theorem* (in a fairly weak form) shows that the group $\overline{G}(R)$ does *not* have *PSG*: indeed the rate of growth of the *congruence subgroups* in $\overline{G}(R)$ is faster than polynomial.

Steps 2–4 serve to establish the characteristic zero case of

THEOREM 1 [**LM3**]. *A finitely generated linear group with PSG is soluble-by-finite.*

The prime-characteristic case of this result follows in an interestingly roundabout manner: if G is a finitely generated linear group over a field of characteristic p, then G is virtually (that is, up to finite index) residually a finite p-group; if G also has *PSG*, one then applies Step 1 to obtain a faithful linear representation of G in characteristic zero.

The most general result along these lines, so far, is

THEOREM 2 [**MS**]. *A finitely generated residually finite-soluble group with PSG is a soluble minimax group.*

[†] This is the group consisting of those $x \in GL_d(R)$ such that for each $n \neq 0$ there exists $g(n) \in G$ with $x \equiv g(n) \pmod{n}$.

This prompts the incidental

PROBLEM 2. Is a residually finite, residually soluble group necessarily residually finite-soluble?

The paper [MS] also contains an analysis of the possible *upper composition factors* of a finitely generated group with PSG (that is, the composition factors of finite quotients of the group). Using also the classification of finite simple groups, we manage to deduce the following (rather *ad hoc*) result:

THEOREM 3 [MS]. *Let G be a finitely generated residually finite group with PSG, and suppose that G contains no infinite strictly ascending chain of normal subgroups N with G/N residually finite. Then G is virtually soluble-minimax.*

Since it is easily seen that the converse also holds (for finitely generated residually finite G), we thus have a characterisation of finitely generated residually finite soluble minimax groups purely in terms of the lattice of finite-index subgroups, labelled with their indices.

Further progress on Problem 1 would seem to require better insight into the possible upper composition factors of a finitely generated group. Very little seems to be known about this in general, so I will state

PROBLEM 3. What can be said about the family of finite simple groups which occur as the upper composition factors of a finitely generated group?

Restricting attention to the groups known to have PSG, it is natural to consider

PROBLEM 4. Determine $\alpha(G)$ in terms of the group-theoretic structure of G, when G is a soluble minimax group.

Some weakish estimates, for finitely generated nilpotent groups, were given in [GSS]. The sharpest available result is

THEOREM 4 [MS]. *For any group G,*

$$\alpha(G) \leqslant 2 \operatorname{ur}(G) + 2.$$

Here, $\operatorname{ur}(G) = \sup \{ \operatorname{rk}(\overline{G}) \mid \overline{G} \text{ a finite quotient of } G \}$ is the *upper rank* of G (where $\operatorname{rk}(\overline{G})$ denotes the rank of $\overline{G}$ in the sense of Prüfer, that is, the least d such that every subgroup of $\overline{G}$ can be generated by d elements).

2. ZETA FUNCTIONS AND POINCARÉ SERIES

Write $a_n(G)$ for the number of subgroups of index exactly n in G (so $a_n(G) = s_n(G) - s_{n-1}(G)$). This is finite if G is finitely generated. We have discussed the growth of this function in §1; what about its arithmetical properties? Following the number theorists, let us consider the generating function

$$\zeta_G(s) = \sum_{n=1}^{\infty} a_n(G)n^{-s} = \sum_{H \leqslant_f G} |G : H|^{-s}$$

where $H \leqslant_f G$ means that H is a subgroup of finite index in G, and s denotes a complex variable. It turns out that the abscissa of convergence of this Dirichlet series is just $\alpha(G)$; so $\zeta_G(S)$ is a reasonable function to consider whenever G has PSG.

A detailed study of $\zeta_G(S)$ for T-groups G, that is, for finitely generated torsion-free nilpotent groups, was begun by Fritz Grunewald, Geoff Smith and myself in [GSS]. (We also considered similar functions which 'count' *normal* subgroups, and subgroups more or less *isomorphic to* G, but for simplicity, I shall ignore these here.) A basic (and easy) fact is the existence of an 'Euler product':

THEOREM 5 [GSS]. *If G is a T-group then*

$$\zeta_G(s) = \prod \zeta_{G,p}(s)$$

where $\zeta_{G,p}(s) = \sum a_{p^n}(G)p^{-ns}$, and the product is over all primes.

This is merely the 'analytic' version of the fact that a finite nilpotent group is the direct product of its Sylow subgroups. Since finite soluble groups also have a very well-controlled Sylow structure (Sylow systems, Hall subgroups), there may be some hope for

PROBLEM 5. Establish a suitable analogue of Theorem 5 for soluble minimax groups G.

(A small hint in this direction is the proof of Proposition 3.3 of [MS].) Returning to T-groups, we see that Theorem 5 now focuses attention on the sequences $a_{p^n}(G)$. The generating function for such a sequence is the *Poincaré series*

$$Z_p(X) = \sum_{n=0}^{\infty} a_{p^n}(G)X^n$$

(to simplify notation we take G as fixed), so

$$\zeta_{G,p}(s) = Z_p(p^{-s}).$$

The main result of [GSS] is that *if G is a T-group then $Z_p(X)$ represents a rational function of X over $\mathbf{Q}$.* To prove it, we re-interpret $Z_p(p^{-s})$ as an integral:

$$(*) \qquad\qquad Z_p(p^{-s}) = \int_M |f(x)|^s |g(x)||d\mu(x)|$$

Here M is a certain p-adic manifold, depending on the structure of G, f and g are certain rational functions, $|\ |$ denotes p-adic absolute value and μ is a natural measure on M. The result is then seen as a special case of a fundamental theorem of J. Denef [D], namely that *an integral of the form $(*)$ is always equal to a rational function of p^{-s}.*

EXAMPLES. (i) If $G = \mathbf{Z}^d$, then $Z_p(X) = (1-X)^{-1}(1-pX)^{-1} \cdots (1-p^{d-1}X)^{-1}$.

(ii) If G is the two-generator free nilpotent group of class two—the 'discrete Heisenberg group' H—then

$$Z_p(X) = \frac{1 - p^3 X^3}{(1-X)(1-pX)(1-p^2X^2)(1-p^3X^2)} \, .$$

With the Euler product these give

$$\zeta_{\mathbf{Z}^d}(s) = \zeta(s)\zeta(s-1)\cdots\zeta(s-d+1),$$

$$\zeta_H(s) = \zeta(s)\zeta(s-1)\zeta(2s-2)\zeta(2s-3) \, / \, \zeta(3s-3)$$

where $\zeta(s)$ is the Riemann zeta function. These and other examples suggest

PROBLEM 6. Prove that if G is a finitely generated relatively free nilpotent group, then there exists a rational function $R(T, X)$ of two variables over $\mathbf{Q}$ such that, for all primes p,

$$Z_p(X) = R(p, X).$$

(This is proved in [GSS] for the free groups in the variety N_2 of class-2 nilpotent groups.)

I am confident that Problem 6 has a positive solution (it is part of a more general conjecture stated in [GSS]). The next suggestion is much more speculative; all the evidence we have supports it, but we have very little evidence:

PROBLEM 7. Prove that if G is a T-group then there exist finitely many rational functions $R_i(T, X)$ over $\mathbf{Q}$ ($i = 1, \ldots, k$) such that for every prime p there exists $i \leqslant k$ with $Z_p(X) = R_i(p, X)$.

The proof of $(*)$ for a T-group G relies on the fact that G has a 'Mal'cev basis', so that—apart from non-commutativity—G rather resembles $\mathbf{Z}^d$. It was therefore astonishing when Marcus du Sautoy established in his thesis that *the rationality of $Z_p(X)$ is a completely general phenomenon*. He proves

THEOREM 6 [dS]. *Let G be a finitely generated group and p a prime. Assume that (i) the sequence $(a_{p^n}(G))_{n \in \mathbf{N}}$ grows at most polynomially, and (ii) the upper p-chief factors of G have bounded order. Then $Z_p(X)$ is a rational function of X.*

This is a vast generalization of the previous result. As well as covering all soluble minimax groups (including the polycyclic groups and T-groups), it covers many non-soluble groups.

Theorem 6 is deduced from the following result, where we write $b_{p^n}(G)$ for the number of *subnormal* subgroups of index p^n in G:

THEOREM 7 [dS]. *Let G be a finitely generated group and p a prime. Assume that the sequence $(b_{p^n}(G))_{n \in \mathbb{N}}$ grows at most polynomially. Then*

$$Z_p^{\triangleleft\triangleleft}(X) = \sum_{n=0}^{\infty} b_{p^n}(G)X^n$$

is a rational function of X.

To establish this, du Sautoy replaces G by its pro-p completion $\widehat{G}_p = \Gamma$, say, and observes that $b_{p^n}(G)$ is just the number of open subgroups of index p^n in Γ. The 'polynomial growth' of this sequence implies that Γ is a p-adic analytic group, by the work of Lubotzky and Mann (see Step 1 in § 1, above), so Theorem 7 follows from

THEOREM 8 [dS]. *Let Γ be a compact p-adic analytic group, having c_n open subgroups of index p^n for each n. Then $\sum c_n X^n$ is a rational function.*

The proof of Theorem 8 is modelled on the original argument for T-groups. As we shall see in § 4 below, a compact p-adic analytic group of dimension d 'rather resembles' $\mathbb{Z}_p^d$, apart from non-commutativity. Using this fact, du Sautoy is able to establish an analogue of the integral formula $(*)$, in which the manifold M and the functions f and g are now 'definable' in a certain language appropriate to the 'locally analytic' theory of the p-adic numbers. The point is now that Denef and van den Dries in [DvD] have recently established the analogue of Denef's rationality theorem, by showing that such an integral is always a rational function of p^{-s}.

Du Sautoy's theorems raise several interesting questions, along the lines of

PROBLEM 8. Let G be an interesting group satisfying the hypotheses of Theorem 7 (for example, $SL_d(\mathbb{Z})$), and let p be a prime.

   (i) Determine $\limsup_{n \to \infty} n^{-1} \log b_{p^n}(G)$.

   (ii) Determine the rational function $Z_p^{\triangleleft\triangleleft}(X)$.

   (iii) How does $Z_p^{\triangleleft\triangleleft}(X)$ vary with p?

3. UNIFORM FINITENESS CONDITIONS

THEOREM 9 [LM2]. *A residually finite group of finite rank is virtually locally soluble.*

(For an infinite group G, the rank $\mathrm{rk}(G)$ of G is the least d such that every finitely generated subgroup of G can be generated by d elements, or ∞ if there is no such d.) This remarkable discovery by Lubotzky and Mann, or rather the ideas underlying it, have opened the way to the study of residually finite groups whose finite quotients satisfy some kind of *uniform finiteness condition*. There are two main steps to their argument. The first depends on finite group theory, in particular the Odd Order Theorem, and establishes

THEOREM 10 [LM2]. *Let G be a group of finite upper rank. Then G has a normal subgroup H of finite index such that every finite quotient of H is soluble.*

The other main step, like Step 1 in §1 above, consists in showing that if H above is finitely generated then its pro-p completion $\widehat{H}_p$ is p-adic analytic, and hence linear, for each prime p. The proof is completed (in the finitely generated case) by showing that a suitable term of the derived series of H embeds into the direct product of finitely many of the $\widehat{H}_p$, and appealing to known results about linear groups of finite rank.

Mann and I, and independently J. S. Wilson, used Theorem 10 together with some of P. Hall's results on finitely generated soluble groups, to obtain a variant of Theorem 9 in which we make a hypothesis about the finite quotients of the group:

THEOREM 11 [MS]. *Let G be a finitely generated residually finite group of finite upper rank. Then G is virtually a soluble minimax group.*

The property of PSG (defined in §1 above) can also be construed as a uniform finiteness condition on finite quotients. For a group G, define

$$\alpha^*(G) = \inf \left\{ \alpha \mid s_n(G) \leqslant n^\alpha \text{ for all } n \right\}.$$

It is easy to see that G has PSG if and only if $\alpha^*(G)$ is finite, and that

$$\alpha^*(G) = \sup \left\{ \alpha^*(\overline{G}) \mid \overline{G} \text{ a finite quotient of } G \right\}.$$

Thus Theorems 2 and 3, in §1, also belong in this section. Indeed they are deduced from Theorems 10 and 11 via

THEOREM 12 [MS]. *There is a function f such that $\operatorname{rk}(H) \leqslant f(\alpha^*(H))$ for every finite soluble group H.*

Another property of a similar kind is what we like to call PIG: *polynomial index growth*. The group G has PIG if $|\overline{G} : \overline{G}^n|$ is bounded above by a polynomial in n, as $\overline{G}$ ranges over all finite quotients of G and n over all positive integers. Of course, if G/G^n is finite for every n (which is the case if G is soluble of finite rank, or— according to Zelmanov's very recent solution of the restricted Burnside problem—if G is finitely generated), this is the same as saying that $|G : G^n|$ grows at most polynomially.

In contrast to the situation with PSG, it is easy to find finitely generated residually finite non-soluble groups with PIG, for example many arithmetic groups satisfying the congruence subgroup property (see [MS], Introduction). It would seem reasonable to formulate two separate problems:

PROBLEM 9. Characterise the residually finite soluble groups with PIG.

PROBLEM 10. For a linear group over $\mathbf{Z}[1/m]$, is PIG equivalent to the congruence subgroup property?

Problem 9 was raised in [S2], and a special case was solved:

THEOREM 13 [S2]. *Let G be a finitely generated residually nilpotent soluble group. Then G has PIG if and only if G is a minimax group.*

The analogue of Problem 4 turned out to have a fairly simple solution. For a group G, I defined

$$\beta(G) = \limsup_{n \to \infty} \frac{\log |G : G^n|}{\log n},$$

and proved

THEOREM 14 [S2]. *For a residually finite soluble minimax group G, $\beta(G)$ is equal to the Hirsch length of G.*

As regards groups which are not necessarily soluble, Mann and I have established the following result (which provides some motivation for Problem 10):

THEOREM 15 [MS]. *A finitely generated residually nilpotent group with PIG is linear over a field of characteristic zero.*

The proof, like that of Theorem 11, proceeds via showing that suitable pro-p completions are p-adic analytic. We were unable to manage without the hypothesis of residual nilpotence in this case, so leaving

PROBLEM 11. Does every finitely generated residually finite group with PIG have a faithful linear representation?

A positive answer to this would settle the 'finitely generated' case of Problem 9, since a finitely generated linear group is virtually residually nilpotent, whereupon Theorem 13 becomes applicable.

4. p-ADIC ANALYTIC GROUPS

Most of the results mentioned in §1 and §3 depend, at least in part, on showing that a group G satisfying some suitable finiteness condition has a faithful linear representation. This is done by showing that the pro-p completion $\widehat{G}_p$ of G is p-adic analytic, the prime p being chosen so that G embeds into $\widehat{G}_p$, that is, so that G is residually a finite p-group. This idea was first used by Alex Lubotzky to solve a famous old problem, namely to give a *group-theoretic characterisation of linear groups* (at least of the finitely generated ones in characteristic zero):

THEOREM 16 [L2]. *Suppose the group G has a descending chain of normal subgroups (N_i) such that, for some fixed prime p and some fixed positive integer d,*

   (i) *$|G : N_1|$ is finite;*

  (ii) *N_1/N_i is a finite p-group for all $i \geqslant 1$;*

 (iii) *N_i/N_j is a d-generator group for all $j \geqslant i \geqslant 1$; and*

 (iv) *$\bigcap N_i = 1$.*

Then G has a faithful linear representation over a field of characteristic zero.

Conversely, if G is a finitely generated linear group over a field of characteristic zero, then G has such a chain (N_i).

More recently, as Step 1 in their proof of Theorem 1 (see §1 above), Lubotzky and Mann proved

THEOREM 17 [LM3]. *The statement of Theorem 16 (both directions) holds with condition (iii) replaced by:*

$$\alpha^*(N_1/N_j) \leqslant d \text{ for all } j \geqslant 1.$$

The 'converse' parts of both these theorems are established by showing that G has a faithful linear representation over $\mathbf{Z}_p$, for some prime p, and using the congruence subgroups modulo powers of p. Of more immediate interest to us here are the 'direct' parts. Replacing G by its finite-index subgroup N_1, we have G embedded in the pro-p group

$$\Gamma = \varprojlim_i G/N_i ;$$

and the heart of the matter is to show that a pro-p group whose quotients by open normal subgroups satisfy a suitable uniform finiteness condition is p-adic analytic. Now a *p-adic analytic group* is simply a Lie group whose local co-ordinate systems are $\mathbf{Q}_p$-valued instead of $\mathbf{R}$-valued. The theory of these groups was developed by Michel Lazard in [L1]; in particular, he established the following group-theoretic criterion: *the pro-p group Γ is p-adic analytic if and only if* (a) Γ *is (topologically) finitely generated and* (b) Γ *has an open normal subgroup Δ such that $\Delta/\overline{\Delta^p}$ (or $\Delta/\overline{\Delta^4}$ if $p = 2$) is abelian.* ($\overline{\Delta^p}$ denotes the closure of Δ^p.) Lubotzky and Mann coined the term *powerful* for the stated condition on Δ, and proved the following result, which explains the relevance of this notion to our problems.

THEOREM 18 [LM1]. *Let Γ be a finitely generated pro-p group. Then Γ has a powerful open normal subgroup if and only if Γ has finite rank.*

Here, *rank* is used in the topological sense: it turns out that for a pro-p group (or indeed for a profinite group) Γ, the rank is equal to

$$\mathrm{rk}(\Gamma) = \sup \{ \mathrm{rk}(\Gamma/N) \mid N \text{ an open normal subgroup of } \Gamma \}$$

(note that each Γ/N is a finite group). Thus if G is any group of finite upper rank, then both $\widehat{G}_p$ and the profinite completion $\widehat{G}$ have finite rank.

With hindsight, it has become apparent that (for the purposes we have been discussing) we can dispense altogether with the theory of p-adic analytic groups as such, and deal directly, instead, with *pro-p groups of finite rank*. A detailed account of this approach may be found in the forthcoming book [DMSS]. To indicate the flavour, let me sketch here how one obtains a linear representation of such a pro-p group - not quite a faithful one, but at least one with abelian kernel.

So let Γ be a finitely generated powerful pro-p group. Define $\Gamma_1 = \Gamma$ and, for $i \geqslant 1$, $\Gamma_{i+1} = \overline{\Gamma_i^p [\Gamma_i, \Gamma]}$; this is the 'lower central p-series' of Γ. Because Γ is powerful, the mapping $x \mapsto x^p$ induces epimorphisms $P_i : \Gamma_i/\Gamma_{i+1} \to \Gamma_{i+1}/\Gamma_{i+2}$. Since each Γ_i/Γ_{i+1} is finite, there exists m such that for all $i \geqslant m$, P_i is actually an *isomorphism*. The group Γ_m is then a *uniformly powerful* pro-p group: it is finitely generated, powerful, and the factors in its lower central p-series are all isomorphic.

Changing notation, let us suppose that Γ itself is uniformly powerful (Lazard uses the term 'p-saturable'). In this case, for each n, the mapping $x \mapsto x^{p^n}$ gives a *bijection* between Γ and Γ_{n+1}; denote its inverse by $x \mapsto x^{p^{-n}}$. We now define a new binary operation $+$ on Γ as follows:

$$x + y = \lim_{n \to \infty} (x^{p^n} y^{p^n})^{p^{-n}} .$$

There is also a natural operation of $\mathbf{Z}_p$ on Γ: if $\lambda = \lim a_n$ is a p-adic integer, where $a_n \in \mathbf{Z}$ for each n, one defines

$$x^\lambda = \lim_{n \to \infty} x^{a_n} .$$

THEOREM 19 [L1]. *If Γ is a d-generator uniformly powerful pro-p group, then (with the operations defined above) Γ is a $\mathbf{Z}_p$-module, and as such is free of rank d.*

The action of $\mathrm{Aut}(\Gamma)$ on this $\mathbf{Z}_p$-module now gives us, free of charge, a faithful representation $\mathrm{Aut}(\Gamma) \to GL_d(\mathbf{Z}_p)$. Hence, via the inner automorphisms, we have a representation $\Gamma \to GL_d(\mathbf{Z}_p)$ with kernel the centre of Γ. The conclusion is

THEOREM 20 [DMSS]. *If Γ is a pro-p group of finite rank then there is a representation $\Gamma \to GL_d(\mathbf{Z}_p)$ with abelian kernel.*

This is sufficient for all the applications mentioned in §1 and §3, where linearity was only required as a step towards proving solubility. The construction of a *faithful* linear representation, as in Theorems 16 and 17, takes rather more work: it depends on the construction of the *Lie algebra* associated to a uniformly powerful group Γ, and an application of Ado's theorem (see [DMSS], Part II).

REFERENCES

[D]   J. Denef, 'The rationality of the Poincaré series associated to the p-adic points on a variety', *Invent. Math.* **77** (1984), 1–23.

[DvD]   J. Denef and L. van den Dries, 'p-adic and real subanalytic sets', *Ann. Math.* **128** (1988), 79–138.

[DMSS] J. D. Dixon, A. Mann, M. du Sautoy and D. Segal, *Analytic pro-p groups*, to appear.

[GSS] F. J. Grunewald, D. Segal and G. C. Smith, 'Subgroups of finite index in nilpotent groups', *Invent. Math.* **93** (1988), 185–223.

[L1] M. Lazard, 'Groups analytiques p-adiques', *Publ. Math. IHES* **26** (1965), 389–603.

[L2] A. Lubotzky, 'A group-theoretic characterization of linear groups', *J. Algebra* **113** (1988), 207–214.

[LM1] A. Lubotzky and A. Mann, 'Powerful p-groups 2: p-adic analytic groups', *J. Algebra* **105** (1987), 506–515.

[LM2] A. Lubotzky and A. Mann, 'Residually finite groups of finite rank', *Math. Proc. Cambridge Philos. Soc.* **106** (1989), 385–388.

[LM3] A. Lubotzky and A. Mann, 'Groups of polynomial subgroup growth', *Invent. Math.* (to appear).

[M] A. Mann, 'Some applications of powerful p-groups', in *Groups—St Andrews 1989* (to appear).

[MS] A. Mann and D. Segal, 'Uniform finiteness conditions in residually finite groups', *Proc. London Math. Soc.* (to appear).

[MVW] C. R. Matthews, L. N. Vaserstein and B. Weisfeiler, 'Congruence properties of Zariski-dense subgroups I', *Proc. London Math. Soc.* **48** (1984), 514–532.

[N] M. Nori, 'On subgroups of $GL_n(F_p)$', *Invent. Math.* **88** (1987), 257–275.

[R] D. J. S. Robinson, 'A note on groups of finite rank', *Compositio Math.* **31** (1969), 240–246.

[dS] M. du Sautoy, 'Finitely generated groups, p-adic analytic groups and Poincaré series', *Bull. Amer. Math. Soc.* (to appear).

[S1] D. Segal, 'Subgroups of finite index in soluble groups, I', in *Groups—St Andrews 1985*, ed. by C. M. Campbell and E. F. Robertson, London Math. Soc. Lecture Note Ser. 121, pp. 307–314 (Cambridge University Press, Cambridge, 1986).

[S2] D. Segal, 'Subgroups of finite index in soluble groups, II', in *Groups—St Andrews 1985*, ed. by C. M. Campbell and E. F. Robertson, London Math. Soc. Lecture Note Ser. 121, pp. 315–319 (Cambridge University Press, Cambridge, 1986).

All Souls College
OXFORD OX1 4AL
Great Britain

MODULAR REPRESENTATIONS OF FINITE GROUPS OF LIE TYPE IN A NON-DEFINING CHARACTERISTIC

To Bernhard Neumann on his eightieth birthday

BHAMA SRINIVASAN

§0. GENERAL THEORY

Let G be a finite group and p a prime. The p-modular representation theory of G was developed by Brauer during the 1930s and 1940s. The idea was to study the representations of G over a field of characteristic p, and the arithmetic of the group ring $\mathfrak{O}G$ where $\mathfrak{O}$ is a discrete valuation ring with maximal ideal $\mathfrak{P}$ such that $\mathfrak{O}/\mathfrak{P}$ is a field of characteristic p. Another objective of the theory was to relate the complex representation theory of G to the p-modular representation theory of G, and then to the p-local structure of G, for various primes p dividing $|G|$.

A survey of Brauer's work can be found in [12]. The basic concepts in modular representation theory of finite groups are described in [6], Chapters 2, 7. A reference for the specific topics mentioned in this section is [6], Chapter 2 §§16-18 and Chapter 7, §56. More advanced and specialized topics are found in [1] and [9].

We fix a p-modular system, i.e. a triple $(K, \mathfrak{O}, k)$ where $\mathfrak{O}$ is a discrete valuation ring in characteristic 0 with quotient field K such that the residue field $k = \mathfrak{O}/\mathfrak{P}$, where $\mathfrak{P}$ is the maximal ideal of $\mathfrak{O}$, is a field of characteristic p. We assume that K is "sufficiently large", e.g. a splitting field for G. It is in fact sufficient to take K to be an algebraic number field containing the $|G|$-th roots of unity, and $\mathfrak{O}$ to be the valuation ring for a $\mathfrak{p}$-adic valuation on K, where $\mathfrak{p}$ is a prime ideal in the ring of integers of K.

The representations of G over K are called ordinary representations and those over k are called p-modular representations. They are related as follows. Let V be a (left) KG-module. Then we can pick an $\mathfrak{O}G$-lattice M in V, that is, an $\mathfrak{O}$-free $\mathfrak{O}G$-module $M \subset V$ such that $M \otimes K = V$. We can then consider $\overline{M} = M \otimes k$, the "reduction mod p" of M, which is a kG-module. Brauer and Nesbitt showed that if M, N are two such $\mathfrak{O}G$-lattices in V then $\overline{M}$ and $\overline{N}$ have the same composition factors.

The Grothendieck group $\mathcal{R}_K(G)$ is a free abelian group with $\mathbb{Z}$-basis the set of isomorphism classes of simple KG-modules, which we denote as $[V_1], [V_2], \ldots, [V_r]$. Similarly we can denote a basis for $\mathcal{R}_k(G)$ by $[L_1], [L_2], \ldots, [L_s]$ where $L_1, L_2, \ldots, L_s$ are

Supported by NSF.

representatives for the isomorphism classes of simple kG-modules. Choosing a lattice M_i in each V_i as above, we get a homomorphism of abelian groups $d: \mathcal{R}_K(G) \to \mathcal{R}_k(G)$ given by $[V_i] \mapsto [\overline{M}_i]$ called the decomposition map. The $r \times s$ matrix which has the multiplicity $(\overline{M}_i : L_j)$ of L_j as a composition factor of $\overline{M}_i$ as its (i,j)-entry is called the p-decomposition matrix of G. This matrix, then, relates the ordinary representation theory of G with the p-modular representation theory of G.

The p-blocks of G also form a central part of Brauer's theory (see [6], Chapter 7). Several definitions are used interchangeably (in spite of the fact that they are not formally equivalent). A central primitive idempotent of the group algebra $\mathfrak{O}G$, or the indecomposable 2-sided ideal of $\mathfrak{O}G$ associated with it, is called a p-block of G. We can also regard a p-block as

(i) a central primitive idempotent of kG, or the indecomposable 2-sided ideal of kG corresponding to the idempotent, or

(ii) a set of modular representations of G, namely the simple kG-modules occurring in the socle of a p-block of G in the sense of (i), or

(iii) a set of ordinary representations (or characters) of G, namely those whose reductions   mod p have irreducible constituents lying in a fixed p-block of G in the sense of (ii).

In this paper we give a survey of recent work on the l-blocks and l-decomposition matrices of a group G which is a finite group of Lie type, where l is a prime which is not equal to the defining characteristic of G.

It is a pleasure to thank Gus Lehrer and the Department of Mathematics, University of Sydney, for their hospitality during my visit in September 1989, and the Australian Research Council for financial support.

§1. THE SYMMETRIC GROUP

As our first example we consider the symmetric group S_n. For any prime p the p-blocks of S_n have been classified, and we will describe the answer below. On the other hand, the p-decomposition matrices are not known except in some special cases. References for this section are a survey article by James [14] for the general theory, and the book by James and Kerber [16] for a proof of the Nakayama conjecture.

The ordinary irreducible representations of S_n are parametrized by partitions of n, and the p-modular irreducible representations of S_n are parametrized by p-regular partitions of n (that is, partitions of n where no p consecutive non-zero parts are equal). Let F be any field. If λ is a partition of n (written $\lambda \vdash n$) then one can construct an FS_n-module S_λ known as a Specht module. If $\operatorname{char} F = 0$ then S_λ is irreducible and this is the classical Specht module. If $\operatorname{char} F = p$ and μ is p-regular then S_μ has a unique irreducible quotient module D_μ; moreover, this D_μ is a composition factor of S_λ only if $\mu \trianglerighteq \lambda$, where $\trianglerighteq$ denotes the usual dominance order of partitions.

Consider a p-modular system $(K, \mathfrak{O}, k)$ as in §0. Then one reduction mod p of a Specht module over K corresponding to $\lambda \vdash n$ is the Specht module S_λ over k. Working now over k, we can state the problem of finding the p-decomposition matrix as: find the composition multiplicities $(S_\lambda : D_\mu)$, where μ is p-regular.

As we stated earlier, the answer to this question is not known. On the other hand, the problem of finding the p-blocks of S_n has an elegant combinatorial solution in terms of p-hooks and p-cores, concepts which were introduced by Nakayama. If $\lambda \vdash n$, consider the Young diagram corresponding to λ. A *hook* of the diagram consists of some node in the diagram along with all the nodes to the right of it and below it. A p-hook is a hook which consists of p nodes. The p-core of the partition λ is the partition corresponding to the diagram obtained by removing p-hooks from the diagram of λ in any order. It can be shown that the p-core is well-defined, that is, it does not depend on the order in which the p-hooks are removed.

Now we consider, for $\lambda \vdash n$, the Specht modules S_λ over the field K of characteristic 0. We then have the following theorem, conjectured by Nakayama and proved first by Brauer and Robinson.

THEOREM. S_λ, S_μ are in the same p-block if and only if λ, μ have the same p-core.

EXAMPLE. If $n = 5$, the S_λ corresponding to $\lambda = \{5\}, \{2^2 1\}$ and $\{21^3\}$ from a 3-block. The partition $\{31^2\}$ is its own 3-core. It corresponds to a block with only one representation in it, or to a 3-block of defect 0 in the sense of Brauer.

§2. Finite Groups of Lie Type: Blocks

A general reference for this section is the book by Carter [5]. The results on blocks of general linear groups and classical groups to be described are in the papers [10] and [11].

Let $\widetilde{G}$ be a connected, reductive algebraic group defined over $\mathbf{F}_q$. Let $F: \widetilde{G} \to \widetilde{G}$ be a Frobenius morphism of $\widetilde{G}$ and let $G = \widetilde{G}^F$ be the finite group of F-fixed points of $\widetilde{G}$. Then G is called a finite group of Lie type. For example, if we took $\widetilde{G} = GL(n, \overline{\mathbf{F}}_q)$ and F the morphism of $\widetilde{G}$ which raises every entry in an $n \times n$ matrix to its qth power, we would get $G = GL(n, q)$. Similarly we would get the classical groups $Sp(2n, q)$, $SO(2n + 1, q)$, $SO^+(2n, q)$ from the corresponding classical groups over $\overline{\mathbf{F}}_q$.

Fix a prime l, where l does not divide q. In our l-modular system we can take $K = \overline{Q}_l$, and k a field of characteristic l. The starting point for the ordinary representations of G is the construction of Deligne and Lusztig. Suppose that $\widetilde{P} = \widetilde{L}\widetilde{V}$

is a parabolic subgroup of $\widetilde{G}$, so that $\widetilde{L}$ (but not necessarily $\widetilde{P}$) is F-stable. Then $\widetilde{L}^F = L$ is called a Levi subgroup of G. If π is a representation of L over $\overline{Q}_l$, using l-adic cohomology one can construct an element $R_L^G(\pi)$ of $\mathcal{R}_K(G)$. Thus we have a map $R_L^G \colon \mathcal{R}_K(L) \to \mathcal{R}_K(G)$. (For a discussion of the properties of this Lusztig map, see [7].) In particular, if $\widetilde{L} = \widetilde{T}$ is an F-stable maximal torus of $\widetilde{G}$ then we get the original Deligne-Lusztig map $R_T^G \colon \mathcal{R}_K(T) \to \mathcal{R}_K(G)$. If the parabolic subgroup $\widetilde{P}$ is also F-stable, we have $P = LV \subset G$. Then $R_L^G(\pi)$ can be realized as an induced representation $\mathrm{Ind}_P^G(\widetilde{\pi})$, where $\widetilde{\pi}$ is the pullback of π to P. This is called Harish-Chandra Induction from L to G.

EXAMPLE. Let $G = GL(n, q)$. Then a Levi subgroup L is isomorphic to a product of groups of the form $GL(m, q^d)$. In particular, a torus T is isomorphic to a product of r_1 copies of F_q^*, r_2 copies of $\mathsf{F}_{q^2}^*$, and so on, where $r_1 + 2r_2 + \cdots = n$.

DEFINITION. The irreducible constituents of the $R_T^G(1)$ where T varies over the tori of G and 1 is the trivial character, are called the *unipotent representations* (or characters) of G.

EXAMPLE. In $G = GL(n, q)$ the only unipotent representations are those contained in the permutation representation on the costs of a Borel subgroup. They are parametrized by partitions of n. For a full discussion of unipotent representations in general, see [5].

We now discuss the l-blocks of G. All the l-blocks of G for G general linear, unitary or classical (with connected center: see below) were classified in [10] and [11]. We give below the classification of unipotent representations into blocks. The general classification involves the "Jordan decomposition" of characters of G given by Lusztig.

THEOREM 1. *Let* $G = GL(n, q)$ *or* $U(n, q)$. *Let* e *be the order of* $\varepsilon q \bmod l$, *where*

$$\varepsilon = \begin{cases} 1, & G = GL(n, q), \\ -1, & G = U(n, q). \end{cases}$$

Then two representations of G *parametrized by partitions* λ, μ *are in the same* l-*block if and only if* λ, μ *have the same* e-*core.*

We now take $\widetilde{G}$ to be a classical group with a connected center. Then G is one of the groups $CSp(2n, q)$, $CSO(2n+1, q)$, $CSO^+(2n, q)$, $CSO^-(2n, q)$. We assume l and q are odd. Let e be the order of $q^2 \bmod l$. If l divides $q^e - 1$ (resp. $q^e + 1$) we say l is a linear prime (resp. unitary prime).

The unipotent representations of G are parametrized by combinational objects called *symbols*. In the case of $CSp(2n, q)$ and $CSO(2n+1, q)$, the symbols are described as follows (for the other cases see [5]). A symbol Λ of rank n is of the form $\Lambda = \binom{A}{B}$ where $A = \{a_1, a_2, \dots\}$, $B = \{b_1, b_2, \dots\}$ are subsets of N, $|A| + |B| = 2m+1$ for some m, and $\sum_i a_i + \sum_i b_i = m^2 + n$. (These are symbols of odd defect; in the other cases

one considers symbols of even defect.) There is an equivalence relation on symbols. We
have $\binom{A}{B} \sim \binom{B}{A}$, and furthermore $\binom{A}{B} \sim \binom{A'}{B'}$ where A', B' are obtained from A, B
respectively by adding 0 and increasing each member by 1. Thus, more precisely, an
equivalence class of symbols is associated with a unipotent representation.

We introduce the operations of adding a t-hook and a t-cohook to symbols. We
say $\mu = \binom{C}{D}$ is obtained from $\Lambda = \binom{A}{B}$ by adding a t-hook (resp. a t-cohook) if μ is
obtained from Λ by adding t to a member of A or B (resp. by adding t to a member
of A or B and moving it to the other side). We can then talk of removing t-hooks and
t-cohooks, and of the t-core or t-cocore of a symbol.

THEOREM 2. *Suppose $l \neq 2$. Let $\widetilde{G}$ be a classical group with a connected center,
where q is odd. Then two unipotent representations of G parametrized by symbols Λ,
μ are in the same l-block if and only if they have the same e-core if l is a linear prime,
and the same e-cocore if l is a unitary prime.*

EXAMPLE. Consider the group $CSp(8, q)$ of type C_4. The 2-cocore of the sym-
bols $\binom{0\ 4}{1}$, $\binom{0\ 3}{2}$, $\binom{0\ 1\ 2}{2\ 3}$, $\binom{0\ 1\ 2\ 3\ 4}{1\ 2}$ is $\binom{0}{1\ 2}$. Thus the unipotent representations
parametrized by these symbols are in the same l-block, where l divides $q^2 + 1$.

We make the following remarks on possible generalizations of these results to arbitrary finite groups of Lie type. The combinatorial descriptions of Theorems 1 and 2 can
be formulated in terms of the map R_L^G. Suppose $G = G_n$ is one of the classical groups
mentioned in Theorem 2, of rank n. Let L be a subgroup of G of the form $G_m T$, where
T is a torus of order $q^t - 1$ or $q^t + 1$. Let ψ be a unipotent character of G_m parametrized
by the symbol μ. Then the irreducible constituents of $R_L^G(\psi \cdot 1)$ are parametrized by
symbols Λ, where Λ is obtained from μ by adding a t-hook (resp. t-cohook) when T
is of order $q^t - 1$ (resp. $q^t + 1$). This is analogous to the Murnaghan-Nakayama formula
in the case of S_n (see the proof of James in [15], where he discusses a "hook-wrapping
operator"). Using this, we can give an alternative description of unipotent blocks, i.e.
blocks which contain unipotent representations. Let $\Phi_e(q)$ denote the eth cyclotomic
polynomial, and let G denote a classical group as above.

DEFINITION. A Levi subgroup L of G is *maximally e-split* if $|L|$ is divisible by
the highest power of $\Phi_e(q)$ dividing $|G|$.

DEFINITION. An irreducible representation π of G is *e-cuspidal* if dim π is divisible
by the highest power of $\Phi_e(q)$ dividing $|G/Z(G)|$ (here $Z(G)$ denotes the center of
G). We remark that the usual notion of "cuspidal representation" coincides with "1-
cuspidal" representation.

PROPOSITION. *Suppose l divides $\Phi_e(q)$ but not $\Phi_f(q)$ for $f < e$. The unipotent l-blocks of G are in bijection with the set of pairs (L, π) where L is a maximally e-split Levi subgroup of G and π is an e-cuspidal unipotent representation of L, modulo G-conjugacy. The unipotent representations in an l-block corresponding to (L, π) are the constituents of $R_L^G(\pi)$.*

EXAMPLE. In the example of $CSp(8, q)$ given earlier, the four unipotent representations are the constituents of $R_L^G(\pi)$ where L is a product of a group isomorphic to $CSp(4, q)$ and a torus of order $q^2 + 1$, and π is obtained by taking the unipotent representation corresponding to $\binom{0}{12}$ on $CSp(4, q)$ and the trivial representation of the torus.

Work is in progress to extend this proposition to all finite groups of Lie type.

§3. FINITE GROUPS OF LIE TYPE: l-DECOMPOSITION MATRIX

We will describe here some results of Dipper and James on the l-decomposition matrix of $GL(n, q)$. A reference for this section is [8], where other references are to be found.

Let $G = GL(n, q)$. Dipper and James show the following.

(i) The part of the l-decomposition matrix D corresponding to unipotent representations can be taken to be a lower unitriangular *square* matrix,

$$
V = \begin{pmatrix}
1 & 0 & \cdots & 0 & 0 \\
* & 1 & \cdots & 0 & 0 \\
\vdots & \vdots & \ddots & \vdots & \vdots \\
* & * & \cdots & 1 & 0 \\
* & * & \cdots & * & 1
\end{pmatrix}.
$$

(ii) Suppose we know the matrices V for groups of the form $GL(m, q^d)$. By taking suitable tensor products of such matrices, we can obtain square matrices $D_1, D_2, \ldots D_r$ such that

$$
D = \begin{pmatrix}
D_1 & 0 & \cdots & 0 \\
0 & D_2 & \cdots & 0 \\
\vdots & \vdots & \ddots & \vdots \\
0 & 0 & \cdots & D_r \\
* & * & \cdots & * \\
\vdots & \vdots & & \vdots \\
* & * & \cdots & *
\end{pmatrix}.
$$

Here the unknown entries $*$ can be determined from the upper part of the matrix by the Littlewood-Richardson rule.

(iii) If l divides $q - 1$, the part of V where the columns are indexed by l-regular partitions gives the l-decomposition matrix of S_n.

Thus, the problem of finding D reduces to the problem of determining V.

There are some explicit results for other groups of Lie type of low rank. In particular, the case of $G_2(q)$ has been treated by Hiss (see [13] and the references given there).

§4. BEYOND BLOCKS: WORK OF BROUÉ ET AL.

The work described here is due to Broué, Michel, Cabanes, Enguehard and others. References for this section are [3], [4] and [19].

In this section we take G to be an arbitrary finite group, as in §0. Let $(K, \mathfrak{O}, k)$ be a p-modular system. After determining the classification of the representations in a p-block, the next step is to try to determine various invariants of a block $B = \mathfrak{O}Ge$, where e is a central primitive idempotent in $\mathfrak{O}G$, such as:

(i) the number of ordinary representations (or characters) in B,

(ii) the number of modular representations (or characters) in B,

(iii) the decomposition numbers,

(iv) the generalized decomposition numbers,

(v) the Cartan invariants

(vi) the heights of ordinary characters in B.

(For a definition of these invariants, especially (iv), (v), and (vi), see [9]).

Let $\mathcal{R}_K(G, e)$ be the Grothendieck group of the category of finitely generated $\mathfrak{O}Ge$-modules. In the spirit of Brauer's approach to describe a p-block by p-local information, one tries to compare the block algebras $\mathfrak{O}Ge$ and $\mathfrak{O}Hg$ where H is, say a p-local subgroup and $\mathfrak{O}Hf$ a suitable p-block of H. In the best possible situation the algebras $\mathfrak{O}Ge$, $\mathfrak{O}Hf$ are Morita equivalent, but this is often too much to hope for. Therefore Broué has initiated the study of the derived category $D^b(\mathfrak{O}Ge)$, the objects of which can be taken to be equivalence classes of bounded complexes of $\mathfrak{O}Ge$-modules. (For a quick approach to derived categories, see [18].) One can then ask if the categories $D^b(\mathfrak{O}Ge)$, $D^b(\mathfrak{O}Hf)$ are equivalent as triangulated categories. If this is the case, Broué has shown that there is a *perfect isometry* (to be defined below) between $\mathcal{R}_K(G, e)$ and $\mathcal{R}_K(H, f)$.

DEFINITION. Let G, H be arbitrary finite groups. A generalized character μ of $G \times H$ is called *perfect* if

(i) $\mu(g, h)/|C_G(g)|$, $\mu(g, h)/|C_H(h)| \in \mathfrak{O}$, for $g \in G$, $h \in H$;

(ii) if $\mu(g, h) \neq 0$, then g is a p'-element if and only if h is a p'-element.

Given such a generalized character μ, we can define a monomorphism

$$I_\mu \colon \mathcal{R}_K(H, f) \to \mathcal{R}_K(G, e)$$

by

$$I_\mu(\beta)(g) = \frac{1}{|H|} \sum_{h \in H} \mu(g, h^{-1}) \beta(h).$$

A perfect isometry from $\mathcal{R}_K(H, f)$ to $\mathcal{R}_K(G, e)$ is then a homomorphism of the type I_μ for some μ which is also an isometry. Such a map preserves various invariants of the blocks $\mathcal{O}Ge$, $\mathcal{O}Hf$ such as:

(i) the numbers of ordinary and modular representations (or characters) in the blocks,

(ii) the heights of ordinary characters in the blocks,

(iii) the elementary divisors of the matrices of generalized decomposition numbers and of the Cartan matrices of the blocks.

Some examples of this are the following. The Lusztig map R_L^G, when G is a finite group of Lie type and L is the *dual* of $C_{G^*}(s)$, where s is an l'-semisimple element of the *dual group* G^* and $C_{G^*}(s)$ is a Levi subgroup of G^*, gives a perfect isometry between certain blocks of L and of G. Broué and Fong have shown that if G is a finite simple group with an abelian Sylow 2-subgroup S and $H = N_G(S)$ then there is a perfect isometry between $\mathcal{R}_K(G, e)$ and $\mathcal{R}_K(H, f)$ where e, f correspond to principal blocks. Enguehard has shown that if $G = H = S_n$, $GL(n, q)$ or $U(n, q)$ and if p is a prime not dividing q in the last two cases, there is a perfect isometry between $\mathcal{R}_K(G, e)$ and $\mathcal{R}_K(H, f)$ where e, f correspond to blocks with the same "weight" (a combinational property).

In general, the question here is how far the derived category $D^b(\mathcal{O}Ge)$ is locally determined. This is a fascinating question on which work has barely begun.

§5. OPEN PROBLEMS AND REMARKS

We mention some open problems below, including some on aspects of the theory that we have not touched on in this paper.

1. The problem of determining the p-decomposition matrix of the symmetric group S_n appears to be a very challenging one. This matrix appears as a special case of an l-decomposition matrix of $GL(n, q)$ (when l divides $q - 1$) or of a decomposition matrix of a q-Schur algebra (see [8]). However, the basic problem remains unsolved. We also mention here the problem of determining l-decomposition matrices for all the groups of Lie type.

2. Michler and his group have been studying the Brauer conjectures and Alperin-McKay conjectures for the finite simple groups (see [17]). These conjectures relate the p-blocks of a finite group with those of p-local subgroups. This project remains to be completed, and a general proof for all finite groups appears to be a dream.

3. Alperin's weight conjecture (see [2]) which gives the number of p-modular representations of a finite group in terms of p-local information, has been checked in some cases (including $GL(n, q)$ for p not dividing q, by Alperin and Fong) but remains open in general.

We end this paper with some remarks of a general philosophical nature. Starting with Brauer's theory and moving to explicit calculations in finite groups of Lie type, we have returned to finite groups in general with the work of Broué. These interactions between abstract finite group representations and the representations of finite groups of Lie type appear to us to parallel the interactions between mathematics and physics. They enrich each theory in turn, and provide us with fascinating problems to work on.

REFERENCES

[1] J.L. Alperin, *Local Representation Theory* (Cambridge University Press, Cambridge, 1986).

[2] J.L. Alperin, 'Weights for finite groups', in *The Arcata conference on representations of finite groups*, ed. by Paul Fong, Proc. Symposia in Pure Math. vol. 47 Part 1, pp. 369–379 (Amer. Math. Soc., Providence, 1987).

[3] M. Broué, 'Blocs isometries parfaites, Catégories dérivées', *C. R. Acad. Sci. Paris Sér I. Math.* 307 (1988), 13-18.

[4] M. Broué, 'Isométries Parfaites, Types de Blocs, Catégories Dérivées' (Preprint, École Normale Supérieure, Paris, 1989).

[5] Roger W. Carter, *Finite Groups of Lie Type: Conjugacy classes and complex characters* (Wiley, New York, 1985).

[6] Charles W. Curtis and Irving Reiner, *Methods of Representation Theory*, Vol I (1981), Vol II (1987) (Wiley, New York).

[7] F. Digne et J. Michel, 'Foncteurs de Lusztig et caractères des groupes lineaires et unitaires sur un corps fini', *J. Algebra* 107 (1987), 217–255.

[8] R. Dipper and G. James, 'The q-Schur algebra', *Proc. London Math. Soc.* 59 (1989), 23–50.

[9] Walter Feit, *The Representation Theory of Finite Groups* (North-Holland, Amsterdam, 1982).

[10] Paul Fong and Bhama Srinivasan, 'The blocks of finite general linear and unitary groups', *Invent. Math.* 69 (1982), 109–153.

[11] P. Fong and B. Srinivasan, 'The blocks of finite classical groups', *J. Reine Angew. Math.* 396 (1989), 122–191.

[12] J.A. Green, 'Richard Dagobert Brauer', *Bull. London Math. Soc.* 10 (1977), 317-342.

[13] G. Hiss, 'The decomposition numbers of $G_2(q)$', *J. Algebra* 120 (1989), 339–360.

[14] Gordon James, 'The representation theory of the symmetric groups', in *The Arcata conference on representations of finite groups*, ed. by Paul Fong, Proc. Symposia in Pure Math. vol. 47 Part 1, pp. 111–126 (Amer. Math. Soc., Providence, 1987).

[15] G.D. James, *The representation theory of the symmetric groups*, Lecture Notes in Math. 682 (Springer-Verlag, Berlin, 1978).

[16] Gordon James and Adalbert Kerber, *The representation theory of the symmetric group*, Encyclopedia Math. Appl. 16 (Addison-Wesley, Reading, Mass., 1981).

[17]   Gerhard O. Michler, 'Modular Representation Theory and the Classification of Finite Simple Groups', in *The Arcata conference on representations of finite groups*, ed. by Paul Fong, Proc. Symposia in Pure Math. vol. 47 Part 1, pp. 223–232 (Amer. Math. Soc., Providence, 1987).

[18]   Leonard L. Scott, 'Simulating algebraic geometry with algebra, I: The algebraic theory of derived categories', in *The Arcata conference on representations of finite groups*, ed. by Paul Fong, Proc. Symposia in Pure Math. vol. 47 Part 2, pp. 271–281 (Amer. Math. Soc., Providence, 1987).

[19]   Michel Broué et Jean Michel, 'Blocs et séries de Lusztig dans un groupe réductif fini', *J. Reine Angew. Math.* **395** (1989), 56–67.

Department of Mathematics
University of Illinois at Chicago
CHICAGO IL 60680
USA

ON THE EFFICIENCY OF SOME DIRECT POWERS OF GROUPS

C. M. CAMPBELL, E. F. ROBERTSON AND P. D. WILLIAMS

1. INTRODUCTION

A finite group G is said to be *efficient* if G has a presentation with d generators and r relators where $r - d$ is the rank of the Schur multiplier $M(G)$ of G. Schur [6] showed that any presentation for G with d generators requires at least $d + \text{rank}(M(G))$ relators. Swan [7] showed that not all finite groups are efficient. The minimum of $r - d$ taken over all finite presentations of G is the *deficiency* of G, denoted by def(G). Hence G is efficient if def$(G) = \text{rank}(M(G))$. For background on questions of efficiency and deficiency zero see, for example, [1], [3] and [5]. We denote the direct power of n copies of a group G by G^n.

In [8] Wiegold asked questions concerning the efficiency of direct powers of groups and, in response to these questions, Kenne [4] proved that $PSL(2,5)^2$ is efficient while the more general result that $PSL(2,p)^2$ is efficient for all primes p is given in [2]. In [1] it is shown that the covering group $SL(2,5)^2$ of $PSL(2,5)^2$ is also efficient, this last result giving an example of a direct square with deficiency zero.

In this paper we give some further results on the efficiency of direct powers. We prove:

THEOREM 1. *Let D_{2m} denote the dihedral group of order $2m$. Then, for any $n, m \geqslant 1$, D_{2m}^n is efficient.*

THEOREM 2. *For any odd m, D_{2m}^2 has a covering group which is efficient and is a finite deficiency zero group.*

THEOREM 3. *The direct cube $PSL(2,5)^3$ is efficient.*

2. DIRECT POWERS

The Schur-Künneth formula applied to direct powers of a group G gives

(1)
$$M(G^n) = M(G^{n-1}) \times M(G) \times (G^{n-1} \otimes G).$$

Now

$$M(D_{2m}) = \begin{cases} 1 & m \text{ odd} \\ C_2 & m \text{ even} \end{cases}$$

so

$$\text{rank}(M(D_{2m})) = \begin{cases} 0 & m \text{ odd} \\ 1 & m \text{ even} \end{cases}$$

and an inductive argument gives

$$\text{rank}(M(D_{2m}^n)) = \begin{cases} \frac{1}{2}n(n-1) & m \text{ odd} \\ n(2n-1) & m \text{ even.} \end{cases}$$

Suppose that m is even. To prove that D_{2m}^n is efficient we exhibit a presentation with $2n$ generators and $2n + n(2n-1) = n(2n+1)$ relators. This is true for $n = 1$. Suppose we have a presentation for D_{2m}^k with $2k$ generators and $k(2k+1)$ relators. Then, using induction, the standard presentation for $D_{2m}^k \times D_{2m}$ has $2k + 2$ generators and $k(2k+1)$ relators from the first factor, 3 relators from the second factor and $4k$ commutators. Hence D_{2m}^{k+1} has a presentation with $2(k+1)$ generators and $(k+1)(2k+3)$ relators as required.

Suppose now that m is odd. To prove that D_{2m}^n is efficient we exhibit a presentation with n generators and $\frac{1}{2}n(n+1)$ relators for $n \geqslant 2$. Suppose that such a presentation exists for D_{2m}^2 and D_{2m}^3. Then consider D_{2m}^n for $n \geqslant 4$ and use induction. If $n = 2k$ then $D_{2m}^n = D_{2m}^k \times D_{2m}^k$ and the standard presentation for $D_{2m}^k \times D_{2m}^k$ has $2k$ generators and $\frac{1}{2}k(k+1)$ relators from the first factor, $\frac{1}{2}k(k+1)$ relators from the second factor and k^2 commutators. Hence we have

$$k(k+1) + k^2 = k(2k+1)$$

relators as required. If $n = 2k + 1$ then $D_{2m}^n = D_{2m}^k \times D_{2m}^{k+1}$ and the standard presentation for $D_{2m}^k \times D_{2m}^{k+1}$ has $2k + 1$ generators and $\frac{1}{2}k(k+1)$ relators from the first factor, $\frac{1}{2}(k+1)(k+2)$ relators from the second factor and $k(k+1)$ commutators. Hence we have

$$\tfrac{1}{2}k(k+1) + \tfrac{1}{2}(k+1)(k+2) + k(k+1) = (2k+1)(k+1)$$

relators as required.

The proof of Theorem 1 will now be complete on showing that, for m odd, D_{2m}^2 has a presentation with 2 generators and 3 relators and that D_{2m}^3 has a presentation with 3 generators and 6 relators. We do this in Section 3 by showing that, for m odd,

$$(2) \qquad D_{2m}^2 = \left\langle x,y \mid x^{2m} = 1, \ (x^m y)^2 = 1, \ y^{2m} = (xy)^2 \right\rangle$$

and

$$(3) \qquad D_{2m}^3 = \left\langle a,x,z \mid a^2 = 1, \ (az)^2 = 1, \ (xz^m)^{2m} = 1, \ [x,a] = 1, \right.$$
$$\left. z^{-1}x^{m-1}z = x^{m+1}, \ x^{-1}z^{1-m}x = z^{m+1} \right\rangle.$$

Notice that Theorem 2 is easily deduced from the presentation (2). For consider

$$G = \langle x, y \mid x^{-m}yx^{m}y = 1, \ y^{2m} = (xy)^2 \rangle.$$

Since $x^{2m} = (x^m y)^2$ we see that $x^{2m} \in Z(G)$, the centre of G. Also $x^2 = 1$ in G/G' so $x^{2m} \in G'$, the derived group of G. Therefore $x^{2m} \in Z(G) \cap G'$ and $G/\langle x^{2m}\rangle = D_{2m}^2$. This shows that G is a stem extension of D_{2m}^2 and, since G has deficiency zero, we have shown that G is a covering group of D_{2m}^2.

As a final remark in this section we note that $PSL(2,5)$ is a perfect group with $M(PSL(2,5)) = C_2$. Therefore the Schur-Künneth formula (1) shows that

$$M(PSL(2,5)^n) = C_2^n.$$

We give a 2-generator 5-relator presentation for $PSL(2,5)^3$ in Section 4 which proves Theorem 3.

3. THE EFFICIENCY OF D_{2m}^2 AND D_{2m}^3 FOR m ODD

D_{2m} has the presentation

$$\langle a, b \mid a^2 = b^m = (ab)^2 = 1 \rangle$$

so

$$D_{2m}^2 = \langle a, b, c, d \mid a^2 = b^m = (ab)^2 = c^2 = d^m = (cd)^2 = [a,c] = [a,d] = 1,$$
$$[b,c] = [b,d] = 1 \rangle.$$

Put $x = ad$, $y = bc$ and notice that, when m is odd,

$$a = x^m, \ d = x^{m+1}, \ c = y^m, \ b = y^{m+1}.$$

Thus

$$D_{2m}^2 = \langle x, y \mid x^{2m} = y^{m(m+1)} = (x^m y^{m+1})^2 = y^{2m} = x^{m(m+1)} = 1,$$
$$(x^{m+1}y^m)^2 = [x^m, y^m] = [x^{m+1}, y^{m+1}] = 1 \rangle.$$

This simplifies immediately to give

(4) $D_{2m}^2 = \langle x, y \mid x^{2m} = y^{2m} = (x^m y)^2 = (xy^m)^2 = [x^m, y^m] = [x^{m+1}, y^{m+1}] = 1 \rangle.$

LEMMA 1. The relations $x^{2m} = y^{2m} = (x^m y)^2 = 1$ imply $[x^m, y^m] = 1$.

PROOF. From $x^m y x^m = y^{-1}$ using $x^{2m} = 1$ we have, for any α,

(5) $x^m y^\alpha x^m = y^{-\alpha}.$

Taking $\alpha = m$ and using $x^{2m} = y^{2m} = 1$ gives the result.

LEMMA 2. *The relation* $(xy)^2 = 1$ *holds in* (4).

PROOF. The relations $(xy^m)^2 = 1$ and $y^{2m} = 1$ imply

$$y^{-m}xy^m = x^{-1}$$

which, raised to the power $m+1$, gives $x^{m+1}y^m = y^m x^{-m-1}$. Now

$$1 = x^{-m-1}y^{-m-1}x^{m+1}y^{m+1} = x^{-m-1}y^{-m-1}y^m x^{-m-1}y$$
$$= x^{-m-1}y^{-1}x^{-m-1}y$$
$$= x^{-m-1}y^{-1}x^{-1}y^{-1}x^m \qquad \text{since } (x^m y)^2 = 1.$$

Thus $(xy)^2 = 1$.

We now have

(6) $\qquad D_{2m}^2 = \langle x, y \mid x^{2m} = y^{2m} = (x^m y)^2 = (xy^m)^2 = [x^{m+1},\ y^{m+1}] = (xy)^2 = 1 \rangle.$

LEMMA 3. *The relations* $x^{2m} = 1$, $y^{2m} = 1$, $(x^m y)^2 = 1$, $(xy)^2 = 1$ *imply*

(i) $\qquad\qquad\qquad\qquad [x^2, y^2] = 1,$

(ii) $\qquad\qquad\qquad\qquad (xy^m)^2 = 1.$

PROOF. First note that the conditions of Lemma 1 hold so (5) is valid.

(i) Now $(xy)^2 = 1$ gives $x^{-m-1}y^{-1}x^{-1}y^{-1}x^m = 1$ which, using (5) with $\alpha = 1$, implies that $x^{-m-1}y^{-1}x^{-1}x^m y = 1$. Thus

(7) $\qquad\qquad\qquad\qquad y^{-1}x^{m-1}y = x^{m+1}.$

Square (7) and use $x^{2m} = 1$ to get $y^{-1}x^{-2}y = x^2$ so $[x^2, y^2] = 1$.

(ii) Now (5) with $\alpha = m - 1$ gives

(8) $\qquad\qquad\qquad\qquad x^{-m}y^{m-1}x^m = y^{-m+1}.$

But m is odd so $m - 1$ is even. Use $[x^2, y^2] = 1$ to reduce (8) to

(9) $\qquad\qquad\qquad\qquad xy^{m-1}x^{-1} = y^{-m+1}.$

Now

$$(xy^m)^2 = xy^{m-1}x^{-1}y^{m-1} \qquad \text{since } (xy)^2 = 1$$
$$= 1 \qquad\qquad\qquad\qquad \text{by (9).}$$

We have now proved that

(10) $\qquad\qquad D_{2m}^2 = \langle x, y \mid x^{2m} = y^{2m} = (xy)^2 = (x^m y)^2 = 1 \rangle$

since $m + 1$ is even and $[x^{m+1},\ y^{m+1}] = 1$ is a consequence of $[x^2, y^2] = 1$.

LEMMA 4. If $G = \langle x, y \mid x^{2m} = (x^m y)^2 = 1, \; y^{2m} = (xy)^2 \rangle$, then $G = D^2_{2m}$.

PROOF. First note that by (10) it is sufficient to prove that $y^{2m} = 1$ in G. Now $G = \langle y, xy \rangle$ so $y^{2m} \in Z(G)$. Also $y^2 = 1$ in G/G' so $y^{2m} \in Z(G) \cap G'$. But $G/\langle y^{2m} \rangle = D^2_{2m}$ by (10) so $y^{2m} \in M(D^2_{2m})$ showing that $y^{4m} = 1$.

However $y^{2m} = (xy)^2$ gives, using $(x^m y)^2 = 1$,

$$(11) \qquad x^{m-1} y^{2m} = x^m yxy = y^{-1} x^{m+1} y.$$

Raise (11) to the power m, noting that y^{2m} is central, to obtain

$$x^{m(m-1)} y^{2m^2} = y^{-1} x^{m(m+1)} y.$$

Since $m - 1$ is even and $x^{2m} = 1$ we have $y^{2m^2} = 1$ which, taken with $y^{4m} = 1$, gives $y^{2m} = 1$ as required.

Our next task is to produce a 3-generator 6-relator presentation for D^3_{2m}, m odd. The standard presentation for $D^2_{2m} \times D_{2m}$ using presentation (2) for D^2_{2m} is

$$(12) \qquad \begin{array}{c} \langle x, y, a, b \mid x^{2m} = (x^m y)^2 = 1, \; (xy)^2 = y^{2m}, \; a^2 = b^m = (ab)^2 = 1, \\ [x, a] = [x, b] = [y, a] = [y, b] = 1 \rangle. \end{array}$$

Put $z = xyb^{-1}$ and eliminate $b = z^{m-1}$, $y = x^{-1} z^m$ to obtain the following presentation for D^3_{2m}:

$$(13) \qquad \begin{array}{c} \langle x, z, a \mid x^{2m} = 1, \; (x^{m-1} z^m)^2 = 1, \; z^{2m} = (x^{-1} z^m)^{2m}, \; a^2 = 1, \; z^{m(m-1)} = 1, \\ (az^{m-1})^2 = 1, \; [x, a] = [x, z^{m-1}] = [a, z^m] = 1 \rangle. \end{array}$$

LEMMA 5. In (13) we have

$$(i) \qquad\qquad\qquad\qquad\qquad z^{2m} \;\; \text{is central},$$

$$(ii) \qquad\qquad\qquad\qquad\qquad z^{2m} = 1.$$

PROOF. (i) Since $[a, z^m] = 1$ we have $[a, z^{2m}] = 1$. Also $z^{2m} = (x^{-1} z^m)^{2m}$ so $[x^{-1} z^m, (x^{-1} z^m)^{2m}] = 1$ gives $[x^{-1} z^m, z^{2m}] = 1$ so $[x, z^{2m}] = 1$.

(ii) From $(x^{m-1} z^m)^2 = 1$ we have, using $[x, z^{m-1}] = 1$,

$$(14) \qquad\qquad\qquad zx^{m-1} z^{-1} = z^{-2m} x^{-m+1}.$$

Raise (14) to the power m and use the facts that $m - 1$ is even and $x^{2m} = 1$ to obtain $z^{2m^2} = 1$. This, together with $z^{m(m-1)} = 1$, gives $z^{2m} = 1$ as required.

We now have D^3_{2m} as

$$(15) \qquad \begin{array}{c} \langle x, z, a \mid x^{2m} = 1, \; z^{2m} = 1, \; (xz^m)^{2m} = 1, \; a^2 = 1, \; [x, a] = 1, \; [a, z^m] = 1, \\ (x^{m-1} z^m)^2 = 1, \; (az^{m-1})^2 = 1, \; [x, z^{m-1}] = 1 \rangle. \end{array}$$

LEMMA 6. *In (15) we have*

(i) $$(x^{m-1}z^m)^2 = 1 \text{ is equivalent to } z^{-1}x^{m-1}z = x^{m+1},$$

(ii) $$(az^{m-1})^2 = 1 \text{ is equivalent to } (az)^2 = 1,$$

(iii) $$[x, z^{m-1}] = 1 \text{ is equivalent to } x^{-1}z^{1-m}x = z^{m+1}.$$

PROOF. (i) Use $[x, z^{m-1}] = 1$ and $z^{2m} = 1$.

(ii) Use $[a, z^m] = 1$ and $a^2 = 1$.

(iii) Use $z^{2m} = 1$.

Using Lemma 6 we have the following presentation for D_{2m}^3:

(16)
$$\langle\, x, z, a \mid (xz^m)^{2m} = 1,\ a^2 = 1,\ (az)^2 = 1,\ [x, a] = 1,\ z^{-1}x^{m-1}z = x^{m+1},$$
$$x^{-1}z^{1-m}x = z^{m+1},\ [a, z^m] = 1,\ x^{2m} = 1,\ z^{2m} = 1 \,\rangle.$$

In order to produce the presentation (3) of D_{2m}^3 we show that the final three relations in presentation (16) are redundant.

LEMMA 7. $[a, z^m] = 1$ *is a consequence of* $(az)^2 = 1$, $a^2 = 1$, $z^{2m} = 1$.

PROOF. Raise $aza = z^{-1}$ to the power m.

LEMMA 8. *In (3) we have*

(i) $$z^{-1}x^2z = x^{-2},$$

(ii) $$x^{2m} = 1,$$

(iii) $$x^{-1}z^2x = z^2,$$

(iv) $$z^{2m} = 1.$$

PROOF. (i) From $z^{-1}x^{m-1}z = x^{m+1}$ we have

(17) $$az^{-1}x^{m-1}za = ax^{m+1}a = x^{m+1}.$$

Use $(az)^2 = 1$ to write (17) as $zax^{m-1}az^{-1} = x^{m+1}$ giving

(18) $$z^{-1}x^{m+1}z = x^{m-1}.$$

Taking the product of (18) with $z^{-1}x^{-m+1}z = x^{-m-1}$ gives (i).

(ii) Since $m + 1$ is even, applying (i) to (18) gives $x^{-m-1} = x^{m-1}$ so $x^{2m} = 1$ as required.

(iii) From $x^{-1}z^{1-m}x = z^{m+1}$ we obtain

$$x^{-2}z^{1-m}x^2 = x^{-1}z^{m+1}x.$$

Apply (i) to $x^{-2}z^{1-m}x^2$, using the fact that $1 - m$ is even, to obtain

(19) $$z^{1-m} = x^{-1}z^{m+1}x.$$

Now taking the product of (19) with $z^{1+m} = x^{-1}z^{-m+1}x$ gives the result.

(iv) Apply (iii) to (19) noting that $m + 1$ is even.

These results complete the proofs of Theorems 1 and 2 given in Section 2.

4. AN EFFICIENT PRESENTATION FOR $PSL(2,5)^3$

An efficient presentation for $PSL(2,p)^2$ is given in [2]. When $p \equiv -1 \pmod 6$ a presentation for $PSL(2,p)^3$ is, using results of [2] and a standard presentation for $PSL(2,p)$,

$$\big\langle\, x,y,a,b \mid x^{3p} = (xy^{-1})^2 = 1,\ (xy^{(-1-p)/2}x^{-1}y^{-4})^2 = y^{-p},$$
$$(yx^{(-1-p)/2}y^{-1}x^{-4})^2 = x^{-p},\ y^{1-p} = x^{-1-p}yx^{-1-p},$$
$$a^2 = b^p = (ab)^3 = (ab^4ab^{(p+1)/2})^2 = 1,$$
$$[x,a] = 1,\ [x,b] = 1,\ [y,a] = 1,\ [y,b] = 1 \,\big\rangle.$$

Let $z = xy^{-1}b^{-1}$, and eliminate $b = z^{p-1}$, $y = z^{-p}x$ to obtain, in the case $p = 5$, the following presentation for $PSL(2,5)^3$:

$$\big\langle\, x,z,a \mid x^{15} = z^{10} = 1,\ (x(x^{-1}z^5)^3 x^{-1}(x^{-1}z^5)^4)^2 = (x^{-1}z^5)^5,$$
$$(z^{-5}x^{-3}z^5x^{-4})^2 = x^{-5},\ (x^{-1}z^5)^4 = x^{-6}z^{-5}x^{-5},\ a^2 = 1,\ z^{20} = 1,$$
$$(az^4)^3 = 1,\ (az^{16}az^{12})^2 = 1,\ [x,a] = 1,\ [x,z^4] = 1,\ [a,z^5] = 1 \,\big\rangle.$$

However, since $[x,a] = 1$ and x, a have coprime orders, the group is generated by z and xa. A presentation for $PSL(2,5)^3$ on z and $w = xa$ is

(20)
$$\big\langle\, z,w \mid z^{10} = 1,\ (z^{-5}w^{12}z^5w^{-4})^2 = w^{10},\ (w^{14}z^5)^4 = w^{-6}z^{-5}w^{10},$$
$$(w^{15}z^4)^3 = 1,\ [w^{16},z^4] = 1,\ [w^{15},z^5] = 1,\ w^{30} = 1,$$
$$(w^{16}(w^{14}z^5)^3w^{14}(w^{14}z^5)^4)^2 = (w^{14}z^5)^5,\ (w^{15}z^6w^{15}z^2)^2 = 1 \,\big\rangle.$$

LEMMA 9. In (20)

(i) $[w^{16},z^4] = 1$ may be replaced by $[w^2,z^2] = 1$,

(ii) $[w^{15},z^5] = 1$ may be replaced by $(w^{15}z^5)^2 = 1$.

PROOF. Use $w^{30} = 1$, $z^{10} = 1$.

LEMMA 10. In (20) the last three relations are redundant.

PROOF. This was checked using a machine implementation of a Todd-Coxeter coset enumeration algorithm.

We have the following presentation for $PSL(2,5)^3$:

$$\big\langle\, z,w \mid z^{10} = 1,\ (z^{-5}w^{12}z^5w^{-4})^2 = w^{10},\ (w^{14}z^5)^4 = w^{-6}z^{-5}w^{10},$$
$$(w^{15}z^4)^3 = 1,\ [w^2,z^2] = 1,\ (w^{15}z^5)^2 = 1 \,\big\rangle.$$

Combine the relations $z^{10} = 1$ and $[w^2,z^2] = 1$ to get the following efficient presentation, again checked by computer:

$$PSL(2,5)^3 = \big\langle\, z,w \mid z^8w^{-2}z^2w^2 = 1,\ (z^{-5}w^{12}z^5w^{-4})^2 = w^{10},$$
$$(w^{14}z^5)^4 = w^{-6}z^{-5}w^{10},\ (w^{15}z^4)^3 = 1,\ (w^{15}z^5)^2 = 1 \,\big\rangle.$$

REFERENCES

[1] C. M. Campbell, T. Kawamata, I. Miyamoto, E. F. Robertson and P. D. Williams, 'Deficiency zero presentations for certain perfect groups', *Proc. Roy. Soc. Edinburgh* **103A** (1986), 63–71.

[2] C. M. Campbell, E. F. Robertson and P. D. Williams, 'Efficient presentations of the groups $PSL(2,p) \times PSL(2,p)$, p prime', *J. London Math. Soc.* (to appear).

[3] D. L. Johnson and E. F. Robertson, 'Finite groups of deficiency zero', in *Homological Group Theory*, ed. by C. T. C. Wall, London Math. Soc. Lecture Note Ser. 36, pp. 275–289 (Cambridge University Press, Cambridge, 1979).

[4] P. E. Kenne, 'Presentations for some direct products of groups', *Bull. Austral. Math. Soc.* **28** (1983), 131–133.

[5] B. H. Neumann, 'Some finite groups with few defining relations', *J. Austral. Math. Soc. Ser. A* **38** (1985), 230–240.

[6] I. Schur, 'Untersuchungen über die Darstellung der endlichen Gruppen durch gebrochene lineare Substitutionen', *J. Reine Angew. Math.* **132** (1907), 85–137.

[7] R. G. Swan, 'Minimal resolutions for finite groups', *Topology* **4** (1965), 193–208.

[8] J. Wiegold, 'The Schur multiplier: an elementary approach', in *Groups—St Andrews 1981*, ed. by C. M. Campbell and E. F. Robertson, London Math. Soc. Lecture Note Ser. 71, pp. 137–154 (Cambridge University Press, Cambridge, 1982).

C. M. Campbell
Mathematical Institute
University of St Andrews
ST ANDREWS, Fife
Scotland

E. F. Robertson
Mathematical Institute
University of St Andrews
ST ANDREWS, Fife
Scotland

P. D. Williams
Department of Mathematics
California State University San Bernardino
SAN BERNARDINO CA 92407
USA

REWRITING SYSTEMS AND HOMOLOGY OF GROUPS

J. R. J. GROVES

1. INTRODUCTION

Anick, in [2], describes a resolution of a field k by free modules over an associative augmented algebra R over k. The construction involves the use of a special type of normal form for the elements of R. The theory of such normal forms and, more particularly, the special form of presentations which lead to them has been expounded by Bergman in [4] for ring theory and is available more generally as the theory of term rewriting systems. The theory has largely been developed within computer science but see, for example, Le Chenadec's book [10] for a more algebraic approach and some specific group-theoretic examples.

When the ring R is the monoid ring kG of a monoid G, the normal form required for Anick's proof can be obtained via a complete rewriting system for G (for definitions see Section 2.1). A very similar idea has been exploited, independently, by Squier [11]. Given a complete rewriting system for a monoid G, Squier constructs an exact complex of length four

$$P_3 \longrightarrow P_2 \longrightarrow P_1 \longrightarrow P_0 \longrightarrow \mathbf{Z}$$

which, in dimension less than or equal to 2, co-incides with the well known relation sequence (see, for example, Section VI.6 of [9]).

The aim of this paper is to describe an alternative approach to the results of Anick (in fact a minor generalisation) which generalises the work of Squier. The general approach is based on that of Squier and retains what I believe is a somewhat more constructive flavour. The general technique can be described as building a 'cubical complex' based on directed graphs associated with the rewriting of certain words in the generators of a complete rewriting system. When, for example, the rewriting system is the natural one which takes as generating set the elements of a group G and as rules all pairs $ab \to c$ with $c = ab$ in G we obtain the bar resolution for G.

It is hoped that the alternative approach used here will enable some more insight into the practical aspects of calculating these resolutions. I was unaware of Anick's work for most of the time I was constructing these arguments and I thank Ken A. Brown (of Glasgow) for bringing Anick's paper to my attention. My original argument gives rise to modules which may in some cases be considerably larger than those produced by Anick's approach. I have included the (relatively minor) alterations necessary to yield the Anick resolution but have left in the original arguments which form a natural generalisation of Squier's proof. Ken S. Brown (of Cornell) has now constructed another proof [7] which has a strong topological flavour.

We give a brief map of the paper. In Section 2 we deal with some preliminaries. We describe the basic terminology of rewriting systems and directed graphs naturally associated with them. We also define the 'cubes' and 'stars' that we use in the proof and furnish the main technical facts we will require concerning them. In Section 3 we give the proof of the main result. This is proved inductively and the technical part of the inductive hypothesis can be found in Section 3.2. There are also two illustrative examples in Section 3.1. In Section 4 we make some comments regarding the technique and then give a number of different examples of the resolution in various special cases.

2. Preliminaries

2.1. Notation and rewriting systems

Throughout the paper, Σ will denote a set (of 'generators') and Σ^* will denote the free monoid (that is, semigroup with identity) on Σ. We will often refer to elements of Σ^* as words in Σ and the identity 1 of Σ^* will correspond to the empty word. A *rewriting system* (Σ, R) on Σ consists of a subset R of $\Sigma^* \times \Sigma^*$ together with Σ. The monoid presented by (Σ, R) is the monoid with presentation on generators Σ and relations all equations $l = r$ with $(l, r) \in R$. If $u, v \in \Sigma^*$ and if $u = alb, v = arb$ with $(l, r) \in R$ and $a, b \in \Sigma^*$, we will write $u \to v$. If there is a sequence

$$u = w_1, \ \ldots, \ w_n = v$$

of words $w_i \in \Sigma^*$ with $w_i \to w_{i+1}$, or $w_i = w_{i+1}$, then we write $u_1 \overset{*}{\to} u_2$ (and say 'u_1 *R*-reduces to u_2'). If, for some u, $u \overset{*}{\to} v$ implies $u = v$, we say that u is *irreducible*.

DEFINITION. *A rewriting system* (Σ, R) *associated with a monoid* G *is complete if the following two conditions hold:*

    (a)   *there are no infinite chains* $u_1 \to u_2 \to \cdots \to u_n \to \cdots$;

    (b)   *each congruence class of the congruence defining* G *as a quotient of* Σ^* *contains exactly one irreducible.*

Thus, in a complete rewriting system, we can rewrite each word in Σ^* to a unique irreducible representing the same element of G and the irreducibles therefore provide a 'normal form' for G. In fact, we shall identify each element of G with its corresponding irreducible and so regard G as a subset of Σ^* (but not, of course, as a submonoid). If $w \in \Sigma^*$, $\overline{w}$ will denote the unique irreducible word in Σ^* which has the same image in G as w.

By a suitable 'tidying' process on (Σ, R), we can also assume that it is minimal in the sense that

(a) if $(l, r) \in R$ then r is irreducible;

(b) if $(l_1, r_1), (l_2, r_2) \in R$ then l_1 is not a subword of l_2.

This minimal rewriting system is uniquely determined by the set of irreducibles. We refer to Groves and Smith [8] or Squier [11] for details.

Given $u, v \in \Sigma^*$ with $u \overset{*}{\to} v$, there will, in general, be many ways of reducing u to v. We will choose, for each such pair u, v, a preferred reduction

$$u = w_1 \to \cdots \to w_n = v.$$

Further, we do this so that the restriction of this preferred reduction to $w_i \to \cdots \to w_j$ is itself the preferred reduction from w_i to w_j. The well-founded nature of the order deriving from $\to$ shows that this is possible. We could, for example, always choose – when a choice is required – the left-most occurrence of (the left hand side of) a rule in each w_i. There will be several occasions in the following when we will invoke this preferred reduction to resolve possible ambiguities.

For further details on rewriting systems in general, we refer to Le Chenadec's book [10].

2.2. DIRECTED GRAPHS AND CUBES

In this sub-section we will establish some terminology. The objects are fairly standard but we have not been able to find a suitable reference.

A *directed graph* Δ consists of a set $V(\Delta)$ of vertices and a set $E(\Delta)$ of edges together with initial and terminal functions $i, t: E(\Delta) \to V(\Delta)$. A *path* in Δ is a sequence

$$p = (e_1, \ldots, e_k) \text{ with } e_j \in E(\Delta), \ t(e_j) = i(e_{j+1}), \ (1 \leqslant j \leqslant k - 1).$$

We extend i, t by defining $i(p) = i(e_1)$, $t(p) = t(e_k)$. (So if we identify an edge with a path of length one, the two possible definitions agree). We also allow, for each $v \in V(\Delta)$, an empty path with no edges and initial and terminal points v. The set of all paths in Δ will be denoted $P(\Delta)$.

We will be concerned only with two types of directed graph.

(1)   Suppose that (Σ, R) is a minimal rewriting system for a monoid G. Then $\Gamma = \Gamma(\Sigma, R)$ is the directed graph with $V(\Gamma) = \Sigma^*$ and $E(\Gamma)$ the set of all instances of a single application of a rewriting rule. More precisely, for each $w \in \Sigma^*$ and each decomposition $w = ulv$ with $(l, r) \in R$ there is an edge e with $i(e) = w = ulv$ and $t(e) = urv$. When necessary we will refer to such an edge as $e = (w, u)$; since we have chosen the rewriting system to be minimal, this notation is unambiguous. If, for

example, $\Sigma = \{a\}$ and $R = \{a^2, 1\}$, then we have three edges with initial vertex a^4, namely $(a^4, 1)$, (a^4, a) and (a^4, a^2).

(2)   For a natural number n, denote the set $\{1, \ldots, n\}$ by $\underline{n}$. Denote the empty set by $\underline{0}$. In either case, denote the power set of $\underline{n}$ by $2^{\underline{n}}$. We can give $2^{\underline{n}}$ the structure of a directed graph – which, abusing notation, will also be denoted $2^{\underline{n}}$. Let $V(2^{\underline{n}}) = 2^{\underline{n}}$ and

$$E(2^{\underline{n}}) = \{(S, x) : S \in 2^{\underline{n}}, \ x \in \underline{n} \setminus S\}.$$

Also define $i((S, x)) = S$, $t((S, x)) = S \cup \{x\}$. The induced ordering is then just the inclusion ordering on subsets. We can identify the vertices and edges of $2^{\underline{n}}$ with the vertices and edges of a standard n-dimensional cube (with one vertex – $\underline{0}$ – 'uppermost'). The notation simply gives a way of referring to this when technical verification is required.

Let Δ be a directed graph (in our applications it will usually be a $\Gamma(\Sigma, R)$). An n-cube μ in Δ consists of a pair of functions

$$\mu_V \colon V(2^{\underline{n}}) \to V(\Delta), \qquad \mu_E \colon E(2^{\underline{n}}) \to P(\Delta)$$

so that if μ_E takes an edge E to a path P, then μ_V takes the initial and terminal points of E to the initial and terminal points of P. Thus μ picks out the 1-skeleton of a geometric cube within the directed graph Δ. (For this reason the reader should be aware that a better but more clumsy name for what I call an n-cube would be '1-skeleton of an n-cube'.) Note that $2^{\underline{n}}$ is an n-cube in itself via the identity mapping. Also a 0-cube in Δ can be identified with a vertex and a 1-cube can be identified with a path.

It will be notationally convenient in the following to drop the subscript notation for the maps associated with μ. Thus we will refer to $\mu(v)$ or $\mu(e)$ for $v \in V(2^{\underline{n}})$ or $e \in E(2^{\underline{n}})$.

2.3. FACES OF CUBES

The resolution we build up in Section 3 will attempt to imitate the chain complex of a cubical complex. Our differential will imitate a genuine boundary map. Since we can identify our cubes with the 1-skeleta of geometric cubes we should clearly define the faces of our cubes so that they are the 1-skeleta of the geometric faces of such cubes. Thus the ideas made explicit below are straightforward even though there is some technical detail required.

For each $i \in \underline{n}$, let τ_i denote the function $\tau_i \colon \underline{n-1} \to \underline{n}$ given by

$$\tau_i(k) = \begin{cases} k & \text{if } k < i, \\ k+1 & \text{if } k \geqslant i. \end{cases}$$

DEFINITION. *Let $i \in \underline{n}$*.

(a) *An upper $(n-1)$-face τ_i^+ of $2^{\underline{n}}$ is the homomorphism $\tau_i^+ : 2^{\underline{n-1}} \to 2^{\underline{n}}$ given by*

$$(\tau_i^+)_V(S) = \tau_i(S) \qquad\qquad (S \in 2^{\underline{n-1}}),$$
$$(\tau_i^+)_E((S,x)) = (\tau_i(S), \tau_i(x)) \qquad (x \in \underline{n-1} \setminus S).$$

(b) *A lower $(n-1)$-face τ_i^- of $2^{\underline{n}}$ is the homomorphism $\tau_i^- : 2^{\underline{n-1}} \to 2^{\underline{n}}$ given by*

$$(\tau_i^-)_V(S) = \tau_i(S) \cup \{i\} \qquad\qquad (S \in 2^{\underline{n-1}}),$$
$$(\tau_i^-)_E((S,x)) = (\tau_i(S) \cup \{i\}, \tau_i(x)) \qquad (x \in \underline{n-1} \setminus S).$$

DEFINITION. *Let ρ be an n-cube in a directed graph Δ. Then an upper $(n-1)$-face ρ_i^+ of ρ is the $(n-1)$-cube $\rho \circ \tau_i^+$. A lower $(n-1)$-face ρ_i^- is the $(n-1)$-cube $\rho \circ \tau_i^-$.*

(The 'composites' here should be interpreted as the pair of maps obtained by taking the composites of the vertex maps and the composites of the edge maps.)

It is easy to verify that the faces of $2^{\underline{n}}$ will correspond with the 1-skeleta of the faces of a geometric cube. Further, the upper faces will be those which include the point $\underline{0}$ and the lower faces those which include the point $\underline{n}$. In particular, when $n = 1$ and a 1-cube is identified with a path, the upper 0-face can be identified with the initial point of this path and the lower 0-face with the terminal point. Note that a 0-cube has no faces.

We now have the beginnings of a 'boundary map' on n-cubes. The next three results will provide some of the technical justification for our final use of this 'boundary map'. They are routine technical exercises and we omit the proofs.

LEMMA 2.1. *Let ρ be an n-cube in a directed graph Δ. If $i \in \underline{n}$, $j \in \underline{n-1}$, $i \leqslant j$, then*

$$(\rho_i^\varepsilon)_j^\eta = (\rho_{j+1}^\eta)_i^\varepsilon \qquad (\varepsilon, \eta \in \{+,-\}).$$

COROLLARY 2.2. *Let $\Omega^{\varepsilon,\eta}$ denote $\{ (\rho_i^\varepsilon)_j^\eta : i \in \underline{n}, \ j \in \underline{n-1}, \ \varepsilon, \eta \in \{+,-\} \}$. Then*

(a) *$\Omega^{\varepsilon,\varepsilon}$ is the union $\bigcup_{i \leqslant j} \{ (\rho_i^\varepsilon)_j^\varepsilon, (\rho_{j+1}^\varepsilon)_i^\varepsilon \}$ of equal pairs;*

(b) *there is a bijective correspondence, with corresponding elements equal, between $\Omega^{\varepsilon,\eta}$ and $\Omega^{\eta,\varepsilon}$ given by*

$$(\rho_i^\varepsilon)_j^\eta \longleftrightarrow (\rho_{j+1}^\eta)_i^\varepsilon \qquad \text{if } i \leqslant j.$$

Although the order of the set $\underline{n}$ should not make a significant difference to our ideas, we cannot ignore it entirely. Thus we need to look at the effect of a change of this order. If ρ is an n-cube in Δ and π is a permutation of $\underline{n}$ (and, by extension, of $2^{\underline{n}}$), then $\rho \circ \pi$ is also an n-cube. We need to investigate the faces of $\rho \circ \pi$. Since, however, the 'adjacent transpositions' of the form $(j, j+1)$ generate the symmetric group S_n, it will suffice to do this in case $\pi = (j, j+1)$.

LEMMA 2.3. *Let π denote the permutation $(j, j+1)$ of $\underline{n}$ and let ρ be an n-cube. Then, for each $i \in \underline{n}$ and $\varepsilon \in \{+, -\}$,*

$$
(\rho \circ \pi)_i^\varepsilon = \begin{cases} \rho_{i+1}^\varepsilon & \text{if } i = j, \\ \rho_i^\varepsilon & \text{if } i = j+1, \\ \rho_i^\varepsilon \circ \pi' & \text{if } i \neq j, j+1. \end{cases}
$$

where π' is the permutation of $\underline{n-1}$ given by

$$
\pi' = \begin{cases} (j, j+1) & \text{if } j+1 < i, \\ (j-1, j) & \text{if } j > i. \end{cases}
$$

The proof is again a straightforward combinatorial exercise and is omitted.

2.4. CUBES IN THE DIRECTED GRAPH OF A REWRITING SYSTEM

Suppose that (Σ, R) is a minimal complete rewriting system for a monoid G and let $\Gamma = \Gamma(\Sigma, R)$ be the corresponding graph. Observe that an n-cube μ in Γ provides a description of different R-reductions – with up to n different starting points – of the word $\mu(\underline{0})$ in Σ^*. We shall frequently want to study only the beginnings of these R-reductions – occurring at the 'top layer' of the cube – and so we introduce a notation for this.

An n-star in Γ ($n \geq 0$) will consist of a vertex w and n edges e_j with $i(e_j) = w$. Within stars we also allow the possibility of an 'empty edge' with initial and terminal points equal to w. (This is strictly an empty *path* but the slight abuse of notation is convenient). If $e_j = (w, a_j)$, we will suppose the e_j ordered so that a_j is a prefix of a_{j+1}. (Empty edges may be placed arbitrarily in the order.) Thus we order from the left by occurrence of the rules. We then denote the n-star by

$$
[w; e_1, \ldots, e_n].
$$

If the star contains an empty edge, we shall say it is *degenerate*.

Thus 0-stars and non-degenerate 1-stars can be identified, respectively, with vertices and edges in Γ .

The fact that the vertex set of Γ is the monoid Σ^* means that we have a natural action of Σ^* on this vertex set – by multiplication on either the right or the left. It is easy to see that this extends to actions of Σ^* on $E(\Gamma)$ and so to actions of Σ^* on the directed graph Γ. From there we can easily define an action of Σ^* on stars (and also on cubes). That is, identifying vertices of Γ with elements of Σ^*, we can define the product of a 0-star with an arbitrary star. The next step is to observe that this can be generalised into a product of arbitrary stars. Thus define

$$[w_1; e_1, \ldots, e_k].[w_2; e_{k+1}, \ldots, e_{k+l}] = [w_1 w_2; e_1 w_2, \ldots, e_k w_2, w_1 e_{k+1}, \ldots, w_1 e_{k+l}].$$

The product is again a star, with edges in correct order as written. If $k = 0$ or $l = 0$ then this product agrees, after suitable identifications, with the action of Σ^* on stars. It is an easy technical verification – using the fact that Σ^* is a free monoid – that any star decomposes uniquely as a product of indecomposable stars.

When such an indecomposable star has neither empty edges nor repeated edges, it will be called *critical*. Thus a critical 0-star is an element of Σ, the critical 1-stars are in 1-1 correspondence with rules in R and the critical 2-stars can be identified with critical pairs (see, for example, [10]).

We can also make a similar definition of product for cubes. In the following, $T - k$ denotes $\{t - k : t \in T\}$.

DEFINITION. *Let σ be a k-cube in Γ and τ an l-cube. Suppose $k + l = n$ and define the product n-cube $\sigma \times \tau$ as follows:*

let $U \subseteq 2^n$ and write $U = S \cup T$ with $S \subseteq \underline{k}$ and $T \subseteq \underline{k+l} = \{k+1, \ldots, n\}$; let $x \in \underline{n} \setminus U$; and

define

$$(\sigma \times \tau)(U) = \sigma(S)\tau(T - k),$$

$$(\sigma \times \tau)((S, x)) = \begin{cases} \sigma((S, x))\tau(T - k) & \text{if } x \in \underline{k}, \\ \sigma(S)\tau((T - k, x)) & \text{if } x \in \underline{k+l}. \end{cases}$$

We omit the routine checks that this does, indeed, define an n-cube. We note that, for $k = 0$ or $l = 0$, this agrees with the action of Σ^* on the edges of Γ. If $\rho = \sigma \times \tau$ with $k, l \geq 1$, then we shall say that ρ is *decomposable*.

There are no surprises in calculating the faces of a product and we again leave the following routine technical verification to the reader.

LEMMA 2.4. *Let ρ_1 be a k-cube and ρ_2 an l-cube with $k + l = n$. Then, for each $i \in \underline{n}$, $\varepsilon \in \{+, -\}$,*

$$(\rho_1 \times \rho_2)_i^\varepsilon = \begin{cases} (\rho_1)_i^\varepsilon \times \rho_2 & \text{if } i \leq k, \\ \rho_1 \times (\rho_2)_{i-k}^\varepsilon & \text{if } i > k. \end{cases}$$

We developed the notion of star by considering the 'top layer' of a cube and so each cube ρ has an associated star obtained by taking the edges emanating from $\rho(\underline{0})$.

Of course we must both supply empty edges and re-order, if necessary. If π is the permutation which gives the necessary re-ordering then we shall call sign(π) the *sign of* ρ. Where there is more than one choice for this sign, we assign it arbitrarily; its value will then not be important.

We now turn to the converse problem of associating an n-cube ρ with an n-star. Suppose, firstly, that $\rho = [w; e_1, \ldots, e_n]$ is indecomposable. Let the edge e_i involve application of a rule for which the left hand side is the subword of w occupying positions $o(i), \ldots, t(i)$. Suppose $S = \{i_1, \ldots, i_m\} \subseteq \underline{n}$ with $i_1 \leqslant \cdots \leqslant i_m$. Now choose $k_1 = 0$, $k_2, \ldots, k_l = m$ so that

$$t(i_j) \geqslant o(i_{j+1}) \text{ for } k_p < j < k_{p+1} \text{ and } t(i_{k_p}) < o(t_{k_p+1}) \text{ for } 1 < p < l.$$

Thus $\{k_1, \ldots, k_l\}$ mark the largest "overlapping blocks" of subwords of w which correspond to edges e_i with $i \in S$. Let

$$w = a_1 u_1 a_2 \ldots u_{l-1} a_l$$

with u_p occupying positions $o(i_{k_p+1}) \ldots t(i_{k_{p+1}})$ in w. Define $\rho(S)$ to be the word

$$\rho(S) = a_1 \overline{u_1} a_2 \ldots \overline{u_{l-1}} a_l$$

in Γ.

If $S \subseteq T$ there will clearly be at least one path in Γ from S to T. Let the path from w to $w_{\{i\}}$ consist of just the edge e_i and choose all the other paths required to define the n-cube by taking the preferred path in each case. In this way we define an n-cube with associated n-star equal to μ. If μ is an arbitrary n-star then it is a product of indecomposable n-stars and we define the associated n-cube to be the product of the associated indecomposable n-cubes. We shall refer to cubes obtained in this way as *canonical*.

Given any non-degenerate cube we can associate with it a star and then with this star a canonical cube. Not surprisingly, the upper faces (but only the upper faces) of a non-degenerate cube and its associated cube are closely related.

LEMMA 2.5. *Let σ be a non-degenerate n-cube with associated canonical n-cube ρ. Then each upper $(n-1)$-face σ_i^+ of σ and ρ_i^+ of ρ is non-degenerate. Further, σ_i^+ and ρ_i^+ have a common associated $(n-1)$-cube.*

PROOF. The first statement is trivial. Also the $(n-1)$-star of σ_i^+ is obtained from the n-star of σ by omitting an edge (and possibly translating). Since σ and ρ then have a common n-star and the associated canonical cube depends only on the n-star, the result follows easily.

2.5. RELATIONSHIP WITH ANICK'S CHAINS

We turn briefly to the question of the relationship of our techniques, and in particular the n-stars, with the n-chains of Anick [2]. Let G be the cyclic group of order k
with generator a and one rule $a^k \to 1$. It is a simple combinatorial exercise to verify
that the number of critical n-stars for this system is $(k-1)^{(n-1)}$ whereas the number
of Anick's n-chains is one! (This, together with related examples, is, however, the worst
case of which we know.) There is clearly some difference and we now try to explain this.

We shall say that a critical n-star $[w; e_1, \ldots, e_n]$ is *non-overlapping* if for no i do
the subwords involved in the edges e_{i-1} and e_{i+1} overlap. More precisely if $e_j = (w, a_j)$
and e_j involves an application of the rule (l_j, r_j) then we require that $a_{i-1} l_{i-1}$ is a
subword of a_{i+1}. Note that the term 'non-overlapping' refers not to adjacent edges/rules
– this would be of no interest – but to ones separated by a third one.

If we are to use these techniques with non-overlapping n-stars we will need to know
that the cube associated with such a star has faces which, if non-degenerate, are also
associated with non-overlapping stars. For the upper faces of a cube this is clear.

Suppose that ρ is an n-cube associated with a non-overlapping n-star. Denote
$\rho(\underline{0})$ by w. Then the non-overlapping property shows that we can decompose w in the
form

$$w = w_1' w_2' u_2 w_3' \ldots w_{n-1}' u_{n-1} w_n'$$

where $w_i' u_i w_{i+1}'$ is the left hand side of the rule associated with the i-th edge $\rho(\underline{0}, \{i\})$.
(Interpret u_1 and u_n as 1.) Consider the lower face ρ_i^- and assume it is non-degenerate.
The 'apex' $\rho_i^-(\underline{0})$ is

$$w_1' \ldots u_{i-1} \overline{w_i' u_i w_{i+1}'} u_{i+1} \ldots w_{n-1}' u_{n-1} w_n'.$$

The edges leading from this apex are the first edges in the paths $\rho(\{i\}, \{i, j\})$ with
$i \neq j$. Thus the only possible 'overlapping' occurs when the two edges correspond either
to the values for j of $\{i - 2, i + 1\}$ or to the values of $\{i - 1, i + 2\}$. But the rules
applied must be separated in the former case by u_{i-1} and in the latter case by u_{i+1}.

Hence we can restrict ourselves to non-overlapping critical stars – and cubes – and
retain most of the previous discussion. This is still not enough, however. In the example
above there are still many non-overlapping critical n-stars. We define a *right-minimal*
non-overlapping n-star inductively. Suppose that $[w; e_1, \ldots, e_n]$ is a non-overlapping
n-star. Let w_1' be the largest prefix which is not involved in the rule corresponding
to the second edge (this agrees with the notation above) and write $w = w_1' v$ and
$e_i = w_1' e_i' (i > 1)$. We say that $[w; e_1, \ldots, e_n]$ is *right minimal* if

(1) $[v; e_2', \ldots, e_n']$ is a right-minimal non-overlapping $(n-1)$-star;
(2)   for no proper suffix a of w_1' and edge e_1' is there a non-overlapping n-star
        of the form $[av; e_1', ae_2', \ldots, ae_n']$.

An inspection of the definition of Anick's n-chains (but reversing the left-right order) should now convince the reader that these are in 1-1 correspondence with our right-minimal non-overlapping critical n-stars. For reasons which are by now becoming fairly obvious we shall abbreviate 'right-minimal non-overlapping critical' by 'special'. Observe that the underlying word of a non-overlapping n-star always contains a suffix which is a special n-star.

3. THE RESOLUTION

3.1. STATEMENT, PRELIMINARIES AND EXAMPLES

We begin with a brief statement of the Theorem although the essential content, being technical in nature, is in the inductive hypotheses described in Section 3.2. Let (Σ, R) be a minimal complete rewriting system for a monoid G. We continue to identify elements of G with R-irreducibles in Σ^*. Let K be a commutative ring (with 1).

Let P'_n be the free K-module with basis the set of all n-stars. Let Q_n be the submodule generated by

(1) all n-stars with an empty path;
(2) all n-stars with a repeated edge;
(3) all n-stars which are the product of a k-star with a l-star with $k, l \geqslant 1$;
(4) all expressions of the form $[u;][w; e_1, \ldots, e_n][v;] - \overline{u}[w; e_1, \ldots, e_n]$.

(For the latter, note that we are using the product defined for stars.)

Let P_n be the quotient module P'_n/Q_n. We shall abuse notation by identifying stars with their images in P_n. Thus a degenerate star, for example, is identified with zero. P_n has a natural structure as a KG-module via (4) above and the natural product of stars. Under this structure it is a free (left) KG-module with basis the critical n-stars. P_1 has a basis in bijective correspondence with the rules of R and P_0 a basis in bijective correspondence with Σ. Define P_{-1} to be the monoid ring KG.

THEOREM 3.1. *There is a KG-resolution of K:*

$$\cdots \longrightarrow P_n \longrightarrow P_{n-1} \longrightarrow \cdots \longrightarrow P_0 \longrightarrow P_{-1} \longrightarrow K.$$

There is, of course, a dimension shift, by 1, in the naming of these modules. It is very convenient for the terminology of the proof, however, to have P_n generated by n-stars rather than $(n-1)$-stars.

The proof of Theorem 3.1 will occupy the remainder of Section 3.

REMARK. We will also consider the case that P_n is generated by all *special* n-stars. We will generally not give the full proof in this case but will indicate the points at which there is substantial divergence from the main argument.

EXAMPLES

There is sufficient technical detail in the proof that it may be helpful to give some simple examples of the ideas we will use. Let $\Sigma = \{a_1, a_2, a_3, a_4\}$ and let $R_{ji} (1 \leqslant i < j \leqslant 4)$ be the rule $a_j a_i \to a_i a_j$. Let $R = \{R_{ji}\}$. Then (Σ, R) presents the free commutative monoid of rank 4 with basis the image of Σ.

One critical 2-star for (Σ, R) is represented by

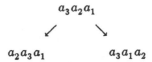

and we can thus produce a (unique) associated 2-cube μ as follows

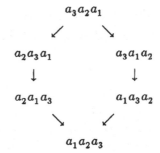

where, for example, the left-hand lower face is given by the unique path from $a_2 a_3 a_1$ to $a_1 a_2 a_3$. This cube has an evident boundary obtained by dividing the faces into single edges. Since, for example, the topmost right-hand edge involves a reduction $a_3 a_2 a_1 \to a_3 a_1 a_2$ we can say that this edge is 'covered' by the element of P_1 corresponding to $a_3 R_{21}$. We then have the following expression, in P_1, for the boundary:

$$a_3 R_{21} + R_{31} + a_1 R_{32} \qquad \text{(right edge)},$$
$$-a R_{32} - a_2 R_{31} - R_{21} \qquad \text{(left edge)}.$$

Note that we have used the left action of Σ^* on G but regard the right action as trivial.

The only indecomposable 3-star ρ for (Σ, R) is the one with underlying word $a_4 a_3 a_2 a_1$ and edges corresponding to the three possible applications of rules. The vertices of the associated cube are as below and the edges between these vertices can be completed in more than one way.

$$\rho(\underline{0}) = a_4 a_3 a_2 a_1$$

$$\rho(\{1\}) = a_3 a_4 a_2 a_1 \qquad \rho(\{2\}) = a_4 a_2 a_3 a_1 \qquad \rho(\{3\}) = a_4 a_3 a_1 a_2$$

$$\rho(\{1,2\}) = a_2 a_3 a_4 a_1 \qquad \rho(\{1,3\}) = a_3 a_4 a_1 a_2 \qquad \rho(\{2,3\}) = a_4 a_1 a_2 a_3$$

$$\rho(\{1,2,3\}) = a_1 a_2 a_3 a_4$$

(The reader is advised to try drawing the full section of the directed graph – it has 24 vertices – below $a_4a_3a_2a_1$ and seeing what choices are possible).

The way to 'cover' faces with critical 2-cubes is to match up the boundary of the face with the boundary of critical 2-cubes. For the upper faces it is clear how to do this; for the lower faces a little more work is needed. For example, the complete graph below the vertex $\rho(\{1\})$ is as follows.

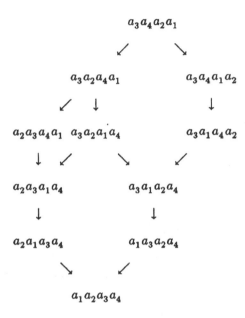

Thus it is reasonable to define a covering for the first lower face to be $a_3[a_4a_2a_1]+[a_3a_2a_1]$ where $[a_ia_ja_k]$ denotes the 2-star corresponding to the word $a_ia_ja_k$.

We are still using the left action of Σ^* and regarding the right action as trivial and we are ignoring the 'square' headed by $a_3a_2a_4a_1$ since it corresponds to a decomposable cube (or star). It is not difficult to verify that, repeating this for all the 6 faces we obtain

face	covering
ρ_1^+	$a_4[a_3a_2a_1]$
ρ_2^+	0
ρ_3^+	$[a_4a_3a_2]$
ρ_1^-	$a_3[a_4a_2a_1] + [a_3a_2a_1]$
ρ_2^-	$a_2[a_4a_3a_1] + [a_4a_2a_1]$
ρ_3^-	$[a_4a_3a_1] + a_1[a_4a_3a_2]$.

We will therefore define the boundary of the 3-star ρ to be the (signed) sum

$$(a_4 - 1)[a_3a_2a_1] - (a_3 - 1)[a_4a_2a_1] + (a_2 - 1)[a_4a_3a_1] - (a_1 - 1)[a_4a_3a_2].$$

The reader familiar with the usual (Koszul complex) resolution for free abelian groups or monoids will recognise this as the expression appropriate to that complex. Full details for this example are given in Section 4.

ORDERINGS

We require some discussion of orderings. There is a natural partial order on Σ^* (or Γ) which defines u to be less than or equal to v if there is a directed path in Γ from v to u or equivalently if $v \xrightarrow{*} u$. Because (Σ, R) is complete, this is a well founded partial order (that is, a partial order with no infinite descending chains). We extend this slightly, by defining, for $u, v \in \Sigma^*$,

$$u \leqslant v \text{ if } u \text{ is a prefix of an } R\text{-reduction of } v.$$

Clearly '$\leqslant$' is still a well-founded partial order on Σ^* and satisfies

$$u \leqslant v \text{ implies } au \leqslant av \text{ for } a \in \Sigma^*.$$

We can further extend '$\leqslant$' firstly to stars and secondly to G-translates of stars (within $\cup_n P_n$) by defining

$$g[v; e_1, \ldots, e_m] \leqslant h[w; f_1, \ldots, f_n] \quad \text{if} \quad gv \leqslant hw.$$

Note that, in this case, '$\leqslant$' need not be antisymmetric and so is a pre-order rather than an order. It is still, however, well founded. Finally, we can extend to the whole of $\cup_n P_n$ by defining, for $\alpha, \beta \in \cup_n P_n$, $\alpha \leqslant \beta$ if each element in the support of α is less than or equal to some element in the support of β. This relation is again a well-founded pre-order.

3.2. THE INDUCTIVE HYPOTHESIS

We fix $n \geqslant 1$ and suppose that, for each $k \leqslant n$, we have the following:
 (a) a KG-module homomorphism $\delta_k \colon P_k \to P_{k-1}$;
 (b) for each k-cube ρ, an element $C_k(\rho)$ of P_k.
These two functions δ_k and C_k should satisfy, for $1 \leqslant k \leqslant n$,
 $(I)_k$ $\delta_{k-1} \circ \delta_k = 0$;
 $(II)_k$ for each $\alpha \in \ker(\delta_{k-1})$, there exists $\beta \in P_k$ with $\beta \leqslant \alpha$ and $\delta_k(\beta) = \alpha$;
 $(III)_k$ given any k-cube ρ,
        (i) if ρ is non-degenerate and is associated with a k-star α then
            $C_k(\rho) - \text{sign}(\rho)\alpha < \rho(\underline{0})$; in any case, $C_k(\rho) \leqslant \rho(\underline{0})$;
        (ii)
$$\delta_k(C_k(\rho)) = \sum_{\substack{i \in \underline{k} \\ \varepsilon \in \{+, -\}}} \varepsilon(-1)^{i+1} C_{k-1}(\rho_i^\varepsilon) \qquad (= D_{k-1}(\rho), \text{ say}).$$

We comment briefly on the significance of these requirements. Firstly $(I)_k$ requires that δ_k be a differential whereas $(II)_k$ requires that the complex be exact with the extra technical requirement that $\beta \leqslant \alpha$. The functions C_k ('C' is for covering) are intended to mimic the idea of a cubical decomposition. So part (i) of $(III)_k$ gives the stars – and their associated cubes – as the basic building blocks of the decomposition. Clearly part (ii) requires that δ_k should imitate the face map on cubes.

We need to begin the induction by defining the maps δ_k, C_k for $k = 0, 1$. The ideas are standard and the material is also well covered in Squier's paper [11]; so we will be brief. The map $P_{-1} \to K$ is, of course, the augmentation map. Recall that a critical 0-star $[a;]$ corresponds to an element a of Σ and a 0-cube to an element of Σ^*. If $a \in \Sigma$, define $\delta_0([a;]) = \bar{a} - 1 \in KG$. If $w = a_1 \ldots a_l \in \Sigma^*$ with $a_i \in \Sigma$, define

$$C_0(w) = \sum_{i=1}^{l} a_1 \ldots a_{i-1}[a_i;].$$

Thus C_0 is essentially the Fox differential.

If $[w; e_1]$ is a critical 1-star, then w is the left hand side of a rewriting rule. Define

$$\delta_1([w; e_1]) = C_0(w) - C_0(\overline{w}).$$

If ρ is a 1-cube, then it corresponds to a path $(e_1, \ldots, e_n)$ in Γ. Suppose that $e_i = (w_i, a_i)$ and that e_i involves the application of a rule (l_i, r_i). Define

$$C_1(\rho) = \sum_{i=1}^{m} a_i[l_i; e_i']$$

where $e_i = a_i e_i'$. It is easily verified that $(III)_1$ is also satisfied. After suitable translation of notation, these definitions agree with the standard ones related to the relation sequence (see, again, the discussion in Squier [11]).

REMARK. In the case that P_n is generated by all *special* n-stars the requirements are similar. We cover only cubes with non-overlapping stars and we cover them with special n-stars. Note that for $n = 0, 1$ there is no difference between 'special' and 'critical'.

3.3. THE INDUCTIVE STEP

INDUCTIVE HYPOTHESIS I

Let $\alpha \in P_{n+1}$ be a critical $(n + 1)$-star and let ρ be the associated canonical $(n + 1)$-cube. Define

$$\delta_{n+1}(\alpha) = D_n(\rho).$$

The verification that $\delta_n \circ \delta_{n+1} = 0$, and so that $(I)_{n+1}$ is satisfied, is now straightforward. In fact, $\delta_k(D_k(\rho)) = 0$ for any k-cube ρ and any $k \leqslant n$; use part (ii) of

$(III)_k$ to express $\delta_k(D_k(\rho))$ in terms of elements of P_{k-1} and Corollary 2.2 to show that the resulting expression is zero. We will use this fact later.

We can now obtain a more explicit expression for δ_k. Suppose that $2 \leqslant k \leqslant n+1$ and that $\alpha = [w; e_1, \ldots, e_k]$ is a critical k-star. Let $[w; e_1, \ldots, \hat{e}_i, \ldots, e_k]$ be the star obtained from α by omitting edge e_i. If $i \neq 1$ or k this will be either critical or decomposable and in the latter case it corresponds to the zero element of P_k. In case $i = 1$ we can write

$$[w; e_2, \ldots, e_k] = a[w'; e_2', \ldots, e_k']$$

where $w = aw'$ and $e_i = ae_i'$.

LEMMA 3.2. *With the notation above,*

$$\delta_k([w; e_1, \ldots, e_k]) = a[w'; e_2', \ldots, e_k'] + \sum_{i=1}^{k-1}(-1)^i[w; e_1, \ldots, \hat{e}_i, \ldots, e_k] + \beta$$

where $\beta < w$.

PROOF. Let ρ be the k-cube associated with $[w; e_1, \ldots, e_k]$. Then we must have

$$\delta_k([w; e_1, \ldots, e_k]) = D_{k-1}(\rho).$$

Let β denote $\sum_{i \in \underline{k}}(-1)^i C_{k-1}(\rho_i^-)$. Then, by $(III)_{k-1}$,

$$C_{k-1}(\rho_i^-) \leqslant \rho_i^-(\underline{0}) \leqslant \rho(i) < \rho(\underline{0}) = w$$

(recall that ρ is non-degenerate). Hence $\beta < w$.

Also, by Lemma 2.5, ρ_i^+ will have an associated cube which is associated with the relevant $(k-1)$-star (i.e. $a[w'; e_2', \ldots, e_k']$ if $i = 1$ and $[w; e_1, \ldots, \hat{e}_i, \ldots, e_k]$ otherwise). The result now follows easily using $(III)_{k-1}$.

REMARK. In the case that P_n is generated by all *special* n-stars the lemma becomes simpler because the terms of the alternating sum will always be zero.

INDUCTIVE HYPOTHESIS *II*

Consider the following condition on an element α of P_n.

(3-1)   If $u[w; e_1, \ldots, e_n]$ is in the **Z**-support of α and e_1 is associated with the rule (l, r), then every proper prefix of ul is reduced.

LEMMA 3.3. *If α satisfies (3-1) and $\delta_n(\alpha) = 0$ then $\alpha = 0$.*

PROOF. Suppose that $\alpha \neq 0$. Choose $x = u[w; e_1, \ldots, e_n]$ in the support of α in such a way that uw is $\leqslant$-maximal amongst all such elements and that, amongst all elements with the same value of uw, the term u has maximal length.

Retaining the notation of Lemma 3.2, we know that the term $x_1 = ua[w'; e_2', \ldots, e_n']$ occurs in the support of $\delta_{n-1}(x)$ (it cannot cancel because of the description given in Lemma 3.2). By (3-1), ua is reduced and because $\delta_n(\alpha') = 0$, x_1 must also occur in the support of $\delta_n(y)$ for some $y = v[w_1; f_1, \ldots, f_n]$ in P_n which is distinct from x.

Using Lemma 3.2, it follows that we can express $\delta_n(y)$ in the form

$$\delta_n(y) = vb[w_1'; f_1', \ldots, f_n'] + \sum_{i=1}^{n-1} (-1)^i v[w_1; f_1, \ldots, \hat{f_i}, \ldots, f_n] + \gamma$$

where $\gamma < vw_1$. Recalling the choice of x, it is clear from the maximality of the term uw of x that $uw = vw_1$ and so that x_1 does not occur in γ. Also, if x_1 were to be a term in the alternating sum above then we would have $v = ua$, contradicting the maximality of u. Thus x_1 is the 'leading' term above; that is

$$ua[w'; e_2', \ldots, e_n'] = vb[w_1'; f_2', \ldots, f_n'].$$

Hence $ua = vb$, $w' = w_1'$ and $e_i' = f_i'$ $(i = 2, \ldots, n)$.

Suppose that e_1 applies the rule with left-hand-side l and f_1 the rule with left-hand-side m. Because of (3-1), every proper prefix of ul and of vm is reduced. But one of ul and vm is a prefix of the other as $uw = vw_1$. Thus $ul = vm$.

Therefore every proper suffix of l and m is reduced and one must be a suffix of the other. Thus $l = m$. Now it is easy to see that $x = y$, contrary to supposition. The proof of Lemma 3.3 is complete.

It remains to reduce to the case of the previous lemma. We do the bulk of this in the following lemma.

LEMMA 3.4. *Any $\alpha \in P_n$ can be written in the form*

$$\alpha = \alpha' + \delta_{n+1}(\beta)$$

where α' satisfies (3-1) and $\beta \leqslant \alpha$.

PROOF. It clearly suffices to prove this when α is a $\mathbf{Z}$-generating element of P_n: say, $\alpha = u[w; e_1, \ldots, e_n]$.

Suppose that α fails to satisfy (3-1). Amongst all such α choose it so that uw is least under the ordering '$\leqslant$' and amongst all such elements with this least value of uw choose it so that u has minimal length.

Retaining the notation above, there must be decompositions $u = u_0 u_1$ and $l = l_0 l_1$ so that $u_1 l_0$ is the left hand side of a rule in R. Let e_0 be the corresponding edge. Then $[u_1 w; e_0, u_1 e_1, \ldots, u_1 e_n]$ is a critical $(n+1)$-star. Further, by Lemma 3.2, we can write

$$\delta_{n+1}([u_1 w; e_0, u_1 e_1, \ldots, u_1 e_n]) = u_1[w; e_1, \ldots, e_n] + \gamma$$

where $\gamma \leqslant u_1 w$ and the terms of the form $a[v; \ldots]$ in the support of γ which have $av = u_1 w$ must also have $a = 1$.

Thus, multiplying by u_0, we obtain

$$\delta_{n+1}(u_0[u_1 w; e_0, u_1 e_1, \ldots, u_1 e_n]) = u[w; e_1, \ldots, e_n] + u_0\gamma.$$

Note that, writing $\beta = u_0[u_1 w; e_0, u_1 e_1, \ldots, u_1 e_n])$, we have that β is bounded by some element in the support of α (in fact α itself) and so $\beta \leqslant \alpha$. Also, each term $\overline{u_0 a}[v; \ldots]$ in the support of $u_0\gamma$ has either $\overline{u_0 a}v \leqslant u_0 av < uw$ or else has $a = 1$ and in this case $\overline{u_0 a}$ is a proper prefix of u.

Using the minimality of α, it follows that each element in the $\mathbf{Z}$-support of $u_0\gamma$ must satisfy (3-1). The proof of the lemma follows.

It is clear that the required result follows from the two lemmas. If α lies in the kernel of δ_n, then write $\alpha = \alpha' + \delta_{n+1}(\beta)$ as in Lemma 3.4. Then, as $\delta_{n+1}(\beta)$ also lies in the kernel of δ_n by the inductive hypothesis $(I)_{n+1}$, it follows that $\delta_n(\alpha') = 0$. Hence $\alpha' = 0$ by Lemma 3.3 and the result follows.

REMARK. In the case that P_n is generated by all *special* n-stars the proof is again somewhat simpler. The condition (3-1) should be altered to replace "every proper suffix of ul" with "the word uw_1'" (using the notation of Section 2.5). In Lemma 3.3 we do not need to worry about the terms in the alternating sum in applying Lemma 3.2 and the arguments of Anick in Lemma 1.3 of [2] can be used to replace the last paragraph of the proof. Lemma 3.4 goes through much as before except that we must decompose uw_1' in such a way as to ensure that the ensuing critical $(n+1)$-star is right-minimal (and so special).

INDUCTIVE HYPOTHESIS III

We will begin by showing that, for any non-degenerate $(n+1)$-cube ρ and any permutation π of $\underline{n}$, we have

$$(3\text{-}2) \qquad D_n(\rho \circ \pi) - \mathrm{sign}(\pi)D_n(\rho) < \rho(\underline{0}) \qquad (= w, \text{ say}).$$

Observe that, if this is true for two permutations π then it is also true for their product. Hence it will suffice to prove it for π an 'adjacent transposition' of the form $(j, j+1)$ – since such transpositions generate the symmetric group. We therefore suppose that $\pi = (j, j+1)$.

Note that

$$C_n(\rho_i^-) \leqslant \rho_i^-(\underline{0}) = \rho(\{i\}) < \rho(\underline{0}) = w.$$

Thus we can, in the expression for D_n, ignore the lower faces of the relevant cubes.

We can now apply Lemma 2.3. Let α_i be the star of ρ_i^+. If $i \neq j, j+1$, then by Lemma 2.3 we have that $(\rho \circ \pi)_i^+ = (\rho_i^+) \circ \pi'$. By part (i) of the inductive assumption $(III)_n$ we also have that

$$C_n(\rho_i^+ \circ \pi') - \operatorname{sign}(\rho_i^+ \circ \pi')\alpha_i < w$$

and

$$C_n(\rho_i^+) - \operatorname{sign}(\rho_i^+)\alpha_i < w.$$

Noting that $\operatorname{sign}(\pi') = \operatorname{sign}(\pi)$, we therefore have that

$$C_n((\rho \circ \pi)_i^+) - \operatorname{sign}(\pi)C_n(\rho_i^+) < w$$

whenever $i \neq j, j+1$.

It is easy to deduce from Lemma 2.3 that the terms for $i = j$ or $i = j+1$ in (3-2) will cancel. Thus the proof of (3-2) follows in the case that ρ is non-degenerate.

We will now show that, if μ is a non-degenerate canonical $(n+1)$-cube which has a star α with zero image in P_{n+1}, then

(3-3) $$D_n(\mu) < \mu(\underline{0}) = w.$$

If α has a repeated edge then the fact that we can have $\mu \circ \pi = \mu$ for an *odd* permutation π, together with (3-2), gives the required result.

So we need to consider (3-3) when α is a product of two stars of strictly smaller size – k and l, say, with $k + l = n + 1$ and $k, l \geqslant 1$. In this case, μ is also such a product. If $k, l > 1$, then Lemma 2.3 shows that each face of μ is a product of this type and so $C_n(\mu_i^\varepsilon) < w$ for all i, ε, by part (i) of $(III)_n$. So (3-3) is true in this case.

Suppose that $\mu = \nu \times \kappa$ with ν a 1-star (or element of Σ^*) and κ an n-star. Then the faces of μ are again decomposable as a product of factors of non-zero dimension except for the pair of faces $\nu_1^+ \times \kappa$ and $\nu_1^- \times \kappa$. But, using part (i) of $(III)_n$ and denoting the star of κ by γ, we have that

$$(\nu_1^+ \times \kappa) - \overline{\nu(\underline{0})}\gamma < w$$

and

$$(\nu_1^- \times \kappa) - \overline{\nu(\{1\})}\gamma < w.$$

Since $\overline{\nu(\underline{0})} = \overline{\nu(\{1\})}$ it follows easily that (3-3) is true in this case also.

Let α be the (image in P_{n+1} of the) star of a $(n+1)$-cube ρ, let π be a permutation that orders the initial edges of ρ into the same order as α and let μ be the cube associated with α.

Suppose that ρ is non-degenerate. We claim that

(3-4) $$D_n(\rho) - \operatorname{sign}(\rho)\delta_{n+1}(\alpha) < \rho(\underline{0}) = w.$$

If α is zero this follows from the above discussion. Otherwise, we can use (3-2) to assume that the leading edges of ρ are ordered in the order in which they occur in α. We can then use Lemma 2.5 to show that $D_n(\rho) - D_n(\mu) < \rho(\underline{0}) = w$.

Finally, we can write $\alpha = u\beta v$ with $u, v \in \Sigma^*$ and β critical. There is then a corresponding decomposition $\mu = u\nu v$ of canonical cubes. Since, by construction, we have $\delta_{n+1}(\beta) = \nu$ we also have $\delta_{n+1}(\alpha) = D_n(\mu)$. Thus (3-4) follows in this case and so now in all cases.

We are now ready to define the 'covering map' C_{n+1}. If ρ is degenerate then observe that $\delta_n(D_n(\rho)) = 0$ and so, by $(II)_{n+1}$ of the inductive hypothesis, we can find γ with $\delta_{n+1}(\gamma) = D_n(\rho)$ and $\gamma \leqslant D_n(\rho) \leqslant \rho(\underline{0})$. Define $C_{n+1}(\rho) = \gamma$. The requirements of $(III)_{n+1}$ are then satisfied.

Suppose now that ρ is non-degenerate. By (3-4) we know that

$$D_n(\rho) - \operatorname{sign}(\rho)\delta_{n+1}(\alpha) < \rho(\underline{0}) = w.$$

But we also know that

$$\delta_n(D_n(\rho) - \operatorname{sign}(\rho)\delta_{n+1}(\alpha)) = 0.$$

Hence, by the inductive hypothesis $(II)_{n+1}$ (which has already been proved), we can find $\gamma \in P_{n+1}$ with

$$\gamma \leqslant D_n(\rho) - \operatorname{sign}(\rho)\delta_{n+1}(\alpha) < w$$

and

$$\delta_{n+1}(\gamma) = D_n(\rho) - \operatorname{sign}(\rho)\delta_{n+1}(\alpha).$$

Define

$$C_{n+1}(\rho) = \operatorname{sign}(\rho)\alpha + \gamma.$$

The two requirements for hypothesis III follow easily. Firstly,

$$C_{n+1}(\rho) - \operatorname{sign}(\rho)\alpha = \gamma < \rho(\underline{0});$$

and secondly,

$$\delta_{n+1}(C_{n+1}(\rho)) = \delta_{n+1}(\operatorname{sign}(\rho)\alpha + \gamma) = D_n(\rho).$$

This completes the inductive step III and with it the proof of Theorem 3.1.

4. COMMENTS, CALCULATIONS AND EXAMPLES

4.1. MONOIDS WITH A FINITE REWRITING SYSTEM

There is an immediate application of the resolution to groups (more generally monoids) with a finite complete rewriting system. If Σ and R are finite, there can clearly be only finitely many critical n-stars because each one is formed from n occurrences of rules in R. It follows that each P_n is a finitely generated KG-module. When there is such a finitely generated resolution of K, we say that G is of type FP_∞ over K. (In fact FP_∞ over $\mathbb{Z}$ implies FP_∞ over any other K.) Thus we have the following theorem.

THEOREM 4.1 (ANICK, SQUIER). *If G is a monoid (group) with a finite complete rewriting system then G is of type FP_∞.*

The theorem is easily deducible from Anick's work in [2] (although Anick's definition of normal form corresponding to a rewriting system is unnecessarily restrictive). It was made explicit by Squier [11] in the special case where FP_3 replaces FP_∞.

In Groves and Smith [8] it is shown that all soluble groups which are constructible in the sense of Bieri and Baumslag [3] have a finite complete rewriting system. The classes of soluble constructible groups and soluble groups of type FP_∞ are not currently known to differ and, in particular, are known to co-incide for metabelian (2-step soluble) groups. Thus the theorem comes close to yielding a group-theoretical characterisation of soluble groups with a finite complete rewriting system.

It may well be, however, that another productive approach is to consider regular rather than finite rewriting systems. (This should probably require that the set of left hand sides of rules should form a regular language). It is far from clear and possibly of some interest to establish what effect this has on the homological properties of the group or monoid. It is not even clear for example whether the homology groups of such monoids, taken with integer coefficients, are restricted in any way.

4.2. INVERSES

A significant drawback, from the point of group theorists, of both rewriting systems and of this resolution is that it concerns monoids and not groups. Groups can, of course, be considered as a special case of monoids in which elements have inverses but most group theorists, including this author, would prefer to think of them as a separate algebraic type with two operations – of inversion as well as multiplication.

The practical consequences here are that rewriting systems for groups must contain rules of the form $aa^{-1} \to 1$ and $a^{-1}a \to 1$ for each generator a of the group whereas in a group presentation this would be taken care of automatically. (These rules can be avoided, of course, when a has finite order.) It would be very useful to have a form of rewriting for groups in which provision of inverses and their cancellation in rewriting was provided automatically.

The effect on the resolution is that it must usually contain many more generators than seems ideally necessary. Consider the submodules generated by stars all edges of which involve rules which cancel a generator with its inverse. It is a simple matter to check that these form a subcomplex and that the quotient by this subcomplex is exact. Thus we can effectively ignore such stars and will usually do so in the following. This does not seem to solve the problem completely, however, as we have not excluded stars which consist largely, but not entirely, of such edges. It would be good to have some way of removing the redundancies that appear to be still present.

There is one case in which we can solve most of these problems.

LEMMA 4.2. *Let G be a group with a submonoid M having the property that $M.M^{-1} = G$. Then*

(1) $Z \otimes_{ZM} ZG \cong Z$;

(2) ZG *is a flat ZM-module;*

(3) *if $\underline{P} \twoheadrightarrow Z$ is a ZM-free resolution of Z then $\underline{P} \otimes_{ZM} ZG \twoheadrightarrow Z$ is a ZG-free resolution of Z.*

The proof is straightforward and is omitted.

4.3. CALCULATION – AND THE BAR RESOLUTION

In the remainder of this section, it will be convenient to adopt a new terminology. If $\alpha = [w; e_1, \ldots, e_k]$ is a critical n-star then we say that w is n-critical. The terminology is not ideal because w does not necessarily determine α but it will save a lot of space.

It is generally straightforward to list all of the generators of the resolution but it is not so straightforward to practically describe the boundary maps. This is not too surprising in that we can not expect the problem of calculating homology to become suddenly easy. We can use the 'cubical' structure to make some progress, however. In the language of this paper the problem is one of determining a covering for the faces of a canonical n-cube associated with a critical n-star. The upper faces are easily dealt with; each such face is either itself associated with a critical $(n-1)$-star or is a product with factors of non-zero dimension and so has zero covering.

We are left with the problem of covering the lower faces of the n-cube. Let μ be a non-degenerate $(n-1)$-cube and let $w = \mu(\underline{0})$. We are going to cover μ with critical $(n-1)$-stars and so we need to 'decompose' μ into the associated $(n-1)$-cubes. For each such cube its 'apex' will include a $(n-1)$-critical word. Thus, in seeking to find a covering of this cube we should look among those words which both include subwords underlying critical $(n-1)$-stars and which are R-reductions of w.

It is not, of course, sufficient to simply add the (appropriately weighted) critical $(n-1)$-stars whose underlying words are R-reductions of w. Consider, for example, the (complete) rewriting system (Σ, R) with

$$\Sigma = \{\, a, b, c, d, e, f \,\},$$

$$R = \{\, ba \rightarrow abf, \ ca \rightarrow ace, \ cb \rightarrow bcd,$$
$$\qquad da \rightarrow ad, \ ea \rightarrow ae, \ fa \rightarrow af, \ db \rightarrow bd, \ eb \rightarrow be, \ fb \rightarrow bf,$$
$$\qquad dc \rightarrow cd, \ ec \rightarrow ce, \ fc \rightarrow cf, \ ed \rightarrow de, \ fd \rightarrow df, \ fe \rightarrow ef \,\}$$

(derived from a subsemigroup of a free nilpotent group of rank three and class two).

Let $n = 2$. There is a 2-critical word cba. Drawing the ordered graph of all reductions of cba it can be seen that there is more than one choice for a covering for the lower face (path) below $bcda$ but the choice is available because there is a word $abcfed$ which contains a 2-critical subword fed. (There are also two words underlying decomposable stars.) The lower face (path) below $cabf$ has no such words and consequently a unique choice of covering.

If there were no 2-critical subwords in the R-reductions of cba we would know that any covering for a lower face is unique – for otherwise the two alternative choices of 1-critical words would bound 2-cubes corresponding to 2-critical words. This is general.

LEMMA 4.3. *Let* μ *be a* $(n-1)$*-cube with* $\mu(\underline{0}) = w$. *Suppose that, among the subwords of reductions of* w, *there is no* n*-critical subword. Suppose also that, the* $(n-1)$*-critical subwords of reductions occur in the form* $b_1, \ldots, b_k$ *with* $u_1 = a_1 b_1 c_1, \ldots,$ $u_k = a_k b_k c_k$ *reductions of* w. *Then the covering* $C_{n-1}(\mu)$ *of* μ *is unique and is an expression of the form* $\varepsilon_1 a_1 \alpha_1 + \cdots + \varepsilon_n a_n \alpha_n$ *with* $\varepsilon_i \in \{-1, 0, 1\}$ *and* α_i *the critical star corresponding to* b_i.

PROOF. The latter statement is clear – and covered in the preceding discussion. For the uniqueness, suppose that d_1 and d_2 are two distinct possibilities for $C_{n-1}(\mu)$. Then $\delta_{n-1}(d_1 - d_2) = 0$ (by III of Section 3.2) and so $d_1 - d_2$ is in the image of δ_n. Thus $d_1 - d_2 = \delta_n(e)$ with $e \leqslant d_1 - d_2$ by II of Section 3.2.

Now take a n-critical word u underlying a n-critical star in the support of e. Then u is bounded by some n-critical word corresponding to a n-critical star in the support of $d_1 - d_2$. Thus $u \leqslant w$. Since this contradicts the initial assumptions we have completed the proof.

In using this result, we need to make two choices. Firstly we must choose which $(n-1)$-stars to include in the covering and secondly we must then assign them a sign. It will, in fact, frequently suffice to take the (positive) sum of all such stars but this is not necessarily the case and the choice needs to be checked in each case. We will omit these checks as they involve largely routine computation.

In what follows we will calculate a number of resolutions using these techniques. We will frequently use what is often known as the *standard* rewriting system for a monoid M; that is, we take $\Sigma = M \setminus \{1\}$ and $R = \{ ab \to \overline{ab} : a, b \in \Sigma \}$ where $\overline{ab}$ denotes either the empty word or the element of Σ representing the product of a and b. We begin with the (very familiar) resolution arising from this.

The n-stars have underlying words $m_1 \ldots m_{n+1}$ and the i-th edge involves an application of the rule $m_i m_{i+1} \to \overline{m_i m_{i+1}}$. Let μ be such a star with associated cube ρ. Observe that the lower faces of ρ have apex of word length n. As the rules are length reducing, no n-critical words, and only the apex itself among $(n-1)$-critical words can occur amongst the reductions of these apices. Hence the i-th lower face, with apex $m_1 \ldots \overline{m_i m_{i+1}} \ldots m_{n+1}$ is covered by the $(n-1)$-star with the same underlying word.

The star corresponding to the word $m_1 \ldots m_{n+1}$ will be written as $[m_1 | \ldots | m_{n+1}]$. (It will be convenient to extend this so that we allow the possibilty that $m_i = 1$ in which case the corresponding star is zero.) Then we have

$$
\begin{aligned}
\delta_n([m_1| \ldots |m_{n+1}]) = {} & m_1[m_2| \ldots |m_{n+1}] && \text{(from the first upper face)} \\
& + (-1)^{n+1}[m_1| \ldots |m_n] && \text{(from the n-th upper face)} \\
& + \sum_{i=1}^{n} (-1)^i [m_1| \ldots \overline{m_i m_{i+1}} \ldots |m_{n+1}] && \text{(from the lower faces).}
\end{aligned}
$$

We thus recover the (normalised) bar resolution for a group or monoid. Had we allowed M, rather than $M \setminus \{1\}$, as the generating set, we would have obtained the unnormalised resolution.

4.4. FREE AND DIRECT PRODUCTS

If G and H are groups with complete rewriting systems, we can form a complete rewriting system for their free product $G * H$ by combining the generating sets – Σ_G and Σ_H say – and combining the sets of rules. Thus a rule of the combined system involves generators from one only of Σ_G and Σ_H. The same therefore follows for critical n-stars and, of course, their boundaries. Thus the resolution $\underline{P} \twoheadrightarrow \mathbf{Z}$ is the direct sum of two resolutions $\underline{P}_G \twoheadrightarrow \mathbf{Z}$ and $\underline{P}_H \twoheadrightarrow \mathbf{Z}$. Here $\underline{P}_G \twoheadrightarrow \mathbf{Z}$ – for example – is the tensor product, over $\mathbf{Z}G$ and with $\mathbf{Z}(G * H)$, of the $\mathbf{Z}G$-free resolution corresponding to the rewriting system for G. The usual facts on the homology of $G * H$ (see, for example p. 220 of Hilton and Stammbach [9]) can be recovered easily.

The direct product $G \times H$ is dealt with similarly. This time, however, we must add a set of rules

$$
R' = \{ hg \to gh : g \in G, h \in H \}.
$$

The n-critical words are thus the juxtapositions of l-critical words in Σ_H (on the left) with m-critical words in Σ_G where $l + m + 1 = n$. (The extra 1 in the sum comes, of course from the rule which interchanges the right-most letter of the H-word with the left-most letter of the G-word.)

The complex obtained is just the tensor product of the complexes obtained from the individual rewriting systems for G and H.

4.5. FREE PRODUCTS WITH AMALGAMATION

The question of rewriting systems for free products with amalgamation (or for HNN-extensions) given rewriting systems for the factors is rather complicated (see [8] for a special case) and we shall not attempt to discuss anything approaching the general case. There is a simple case, however, which we describe briefly.

Let $K = G *_A H$ be the free product of G and H amalgamating A; we shall regard A as a subgroup of G and H. Let S and T be transversals for A in G and H respectively. Let (Σ_A, R_A) be the standard rewriting system for A. Then there is a natural rewriting system for K of the form

$$
\begin{aligned}
ab &\to \overline{ab} & (a, b \in \Sigma_A), \\
s_1 s_2 &\to a(s_1, s_2) s_{12} & (s_1, s_2, s_{12} \in S, a(s_1, s_2) \in \Sigma_A), \\
t_1 t_2 &\to a(t_1, t_2) t_{12} & (t_1, t_2, t_{12} \in T, a(t_1, t_2) \in \Sigma_A), \\
sa &\to b(s, a) u(s, a) & (u(s, a) \in S, b(s, a) \in \Sigma_A), \\
ta &\to b(t, a) u(t, a) & (u(t, a) \in T, b(t, a) \in \Sigma_A).
\end{aligned}
$$

This leads to a resolution which we leave the reader to make explicit but which can be easily described. Let $\underline{P}'_G$ and $\underline{P}'_H$ and $\underline{P}'_A$ be the complexes obtained by taking the rewriting systems for G, H and A obtained from the above. Thus $(\underline{P}'_G)_n$, for example, will be generated by all n-stars with underlying words of the form $s_1 \ldots s_k a_{k+1} \ldots a_{n+1}$.

Let $\underline{P}_G = \underline{P}'_G \otimes_{\mathbf{Z}G} \mathbf{Z}K$ and define $\underline{P}_H$ and $\underline{P}_A$ similarly. Then $\underline{P}_G$ and $\underline{P}_H$ are complexes of $\mathbf{Z}K$-modules which each have a sub-complex isomorphic to $\underline{P}_A$. The $\mathbf{Z}K$-module complex we obtain from the rewriting system above is the direct sum of $\underline{P}_G$ and $\underline{P}_H$ amalgamating the sub-complex $\underline{P}_A$.

This yields a direct, although not particularly elegant, method of deriving the Mayer-Vietoris formula for the homology of an amalgamated free product (cf. Section II.7 of [6].) (Note that it is not easy to give explicit values for the differentials in this discussion – but compare with Section 4.8).

4.6. FREE ABELIAN GROUPS

We will, in fact, deal with free abelian monoids. A free abelian group G clearly contains a free abelian monoid M with $MM^{-1} = G$. So we can apply Lemma 4.2 to recover the group case.

Let A be a free abelian monoid with basis Σ. We shall assume Σ to be a totally ordered set. The set of rules R will be the obvious set

$$R = \{\, ba \to ab : a, b \in \Sigma, a < b \,\}.$$

The n-critical words are then the set of all expressions

$$w = a_1 \ldots a_{n+1} \quad \text{with} \quad a_1 > \cdots > a_{n+1}.$$

Denoting by w_j the effect of applying the j-th rule to w we have

$$w_i = a_1 \ldots a_{i-1} a_{i+1} a_i a_{i+2} \ldots a_{n+1}.$$

Note that the rules do not alter the length of a word; using this and a simple combinatorial argument, it is easy to see that no n-critical word lies among the proper reductions of w. Thus there is a unique form for the covering of the lower faces which we can find by inspecting the $(n-1)$-critical subwords amongst the reductions of the w_j. It is again easy to see that the only such reductions of w_j are

$$a_{i+1}(a_1 \ldots a_{i-1} a_i a_{i+2} \ldots a_{n+1}) \quad \text{and} \quad (a_1 \ldots a_{i-1} a_{i+1} a_{i+2} \ldots a_{n+1}) a_i$$

where the $(n-1)$-critical words have been put in parentheses.

Denote the (unique) star underlying the critical word w by $[w]$. We obtain (after the necessary further checks)

$$
\begin{aligned}
\delta_n(&[a_1 \ldots a_{n+1}]) \\
&= a_1[a_2 \ldots a_{n+1}] && \text{(from the first upper face)} \\
&\quad + (-1)^{n+1}[a_1 \ldots a_n] && \text{(from the n-th upper face)} \\
&\quad + \sum_{i=1}^{n} (-1)^i \big(a_{i+1}[a_1 \ldots a_{i-1} a_i a_{i+2} \ldots a_{n+1}] + [a_1 \ldots a_{i-1} a_{i+1} a_{i+2} \ldots a_{n+1}] \big) \\
& && \text{(from the lower faces)} \\
&= \sum_{i=1}^{n+1} (-1)^{i+1} (a_i - 1)[a_1 \ldots \widehat{a_i} \ldots a_{n+1}].
\end{aligned}
$$

We have obtained the usual complex – often referred to as the *Koszul complex* – for free abelian monoids. (See, for example, Section 6.1 of [3].)

4.7. FINITE CYCLIC GROUPS

The cyclic group case seems the least satisfactory application of this approach.

The approach via 'special stars' (equivalent to that of Anick) yields a complex $\underline{\underline{P}}$ which has one generator in each dimension. More precisely let $G = \langle x : x^k = 1 \rangle$ and let

$$w_n = \begin{cases} x^{ln} & \text{if } k = 2l - 1, \\ x^{ln+1} & \text{if } k = 2l. \end{cases}$$

Then w_n is the unique word underlying a special (see Section 2.5) n-star and the subwords involved in the n edges w_n start at positions $1, n, n+1, 2n, \ldots$. The usual formula for the differential (see for example [6]) can then be deduced.

If, however, we take the 'full' approach using all critical n-stars, then we have many more generators. In fact, we have one critical n-star for each $(n - 1)$-tuple $(l_1, \ldots, l_{n-1})$ with $0 < l_i < k$. Here the corresponding edges will involve subwords which start in positions $1, 1 + l_1, 1 + l_1 + l_2, \ldots$. A little checking will confirm that we have the normalised bar resolution – but shifted by two dimensions.

It seems likely that this problem – of having an excessive number of generators – will recur for any rewriting system which involves elements of finite order.

4.8. EXTENSIONS

Suppose that

$$K \rightarrowtail G \overset{\pi}{\twoheadrightarrow} Q$$

is an extension of groups and that (Σ_K, R_K) and (Σ_Q, R_Q) are complete rewriting systems for K and Q respectively. Then there is a complete rewriting system (Σ_G, R_G) for G with $\Sigma_G = \Sigma_K \cup \Sigma_Q$. Choose a transversal $\tau : Q \to G$ to π – so that $\pi \circ \tau$ is the identity on Q. It will be convenient to regard Σ_Q as a subset of both Q and G so that τ is the identity on Σ_Q.

The rules R_G are then formed from three types:

(1) R_K;

(2) rules of the form $l \to rk$ where $l \to r$ is in R_Q and $\tau(r)^{-1}\tau(l) = k \in \Sigma_K^*$;

(3) rules $ks \to sk'$ where $s \in \Sigma_Q$, $k \in \Sigma_K$, and k' is the R_K-reduced form of the element $s^{-1}ks \in K$.

(See, for example, Groves and Smith [8].)

There is then a corresponding resolution.

We shall elaborate only in the case that (Σ_K, R_K) and (Σ_Q, R_Q) are the standard rewriting systems. We denote the term $\tau(\overline{s_1 s_2})^{-1}\tau(s_1 s_2)$ occurring in rules of type (2) by $u(s_1, s_2)$.

The n-critical words are then the words of the form $w = k_1 \ldots k_p s_1 \ldots s_q$ with $p + q - 1 = n$. The word obtained by applying the i-th rule is

$$
w_i = \begin{cases}
k_1 \ldots \overline{k_i k_{i+1}} \ldots k_p s_1 \ldots s_q & \text{if } i < p, \\
k_1 \ldots k_{l-1} s_1 k_p^{s_1} s_2 \ldots s_q & \text{if } i = p, \\
k_1 \ldots k_l s_1 \ldots \overline{s_j s_{j+1}} u(s_j, s_{j+1}) \ldots s_q & \text{if } i > p, \ j = i - p.
\end{cases}
$$

(Here $\overline{s_j s_{j+1}}$ denotes $\tau \circ \pi(s_j s_{j+1})$. As usual we shall denote the star corresponding to a word by enclosing the word in brackets).

It is easy to check that no n-critical word occurs in the proper reductions of w. (Note that the length of w is never increased by the application of a rule and is decreased unless $i = p$.) In order to find the diferential we would need to find the $(n-1)$-critical words which occur. The problem seems much harder than in previous examples perhaps because there are two problems that have not occurred previously. Firstly, there is a possibility of an $(n-1)$-critical word occurring in two distinct lower faces and secondly such words may also occur in the cover with a negative sign.

It seems likely, however, that this resolution co-incides with one given by André in [1] and this enables us to make an intelligent guess at the appropriate coverings. The covering of the upper faces is, as usual, immediate and the covering of the i'th lower face when $i < p$ can easily be seen to be $[k_1 \ldots \overline{k_i k_{i+1}} \ldots k_p s_1 \ldots s_q]$.

To describe our guess for the coverings of the other lower faces we turn to André's notation in [1]. To understand the next two paragraphs the reader should refer to [1] and the notation described there. We will cover the faces by stars which correspond to the summands of $\delta_n([k_1 \ldots k_p s_1 \ldots s_q])$ with $i \geq 1$. (These stars will also need to be assigned appropriate signs which we will not make explicit but which can be deduced from [1]).

Each such summand contains an entry which is a conjugate of k_p and we will assign the entry to a face according to the precise nature of this conjugate. If the entry is k_p itself, then the star also contains a unique subword $\overline{s_j s_{j+1}} u(s_j, s_{j+1})$; we assign the star to the covering of the $(p+j)$'th lower face. In general the entry is of the form

$$
\rho_\beta^j(s_1, \ldots, s_k)^{-1} k_p \rho_\beta^j(s_1, \ldots, s_k).
$$

Assign the corresponding star to the p'th face if $\beta = \lambda$ and otherwise assign it to the $p + j_{\lambda-1}$'th face.

CONJECTURE. *The above assignment, together with a suitable choice of signs, yields a covering for the faces of the canonical cube associated with $[k_1 \ldots k_p s_1 \ldots s_q]$.*

We can be more explicit in the case that the extension splits. For then the transversal τ can be assumed to be a homomorphism and $u(s_1, s_2)$ is always trivial. In this case the covering of the faces is easily expressed. Let $\alpha = [k_1 \ldots k_p s_1 \ldots s_q]$ be a critical n-star. Then

$$k_1[k_2 \ldots k_p s_1 \ldots s_q] \qquad \text{covers upper face 1,}$$
$$[k_1 \ldots k_p s_1 \ldots s_{q-1}] \qquad \text{covers upper face } n,$$
$$[k_1 \ldots \overline{k_i k_{i+1}} \ldots k_p s_1 \ldots s_q] \qquad \text{covers lower face } i \text{ when } i < p,$$
$$s_1[\overline{k_1^{s_1}} \ldots \overline{k_p^{s_1}} s_2 \ldots s_q] + [k_1 \ldots k_{p-1} s_1 \ldots s_q] \qquad \text{covers lower face } p,$$
$$[k_1 \ldots k_p s_1 \ldots \overline{s_{i-p} s_{i+1-p}} \ldots s_q] \qquad \text{covers lower face } i \text{ when } i > p.$$

It is now straightforward to write down an explicit differential for the resolution. This agrees, after suitable translation of notation, with the split extension case of the resolution given by André.

REFERENCES

[1] André, Michel, 'Homologie des extensions de groupes', *C. R. Acad Sci. Paris* **260** (1965), 3820-3823.

[2] Anick, D. J., 'On the homology of associative algebras', *Trans Amer. Math. Soc.* **296** (1986), 641-659.

[3] Baumslag, G., and Bieri, R., 'Constructible solvable groups', *Math. Z.* **151** (1976), 249–257.

[4] Bergman, G. M., 'The diamond lemma for ring theory', *Adv. in Math.* **292** (1978), 178-218.

[5] Blackburn, Norman, 'Some homology groups of wreathe products', *Illinois J. Math.* **16** (1972), 116-129.

[6] Brown, Kenneth S., *Cohomology of groups* (Springer-Verlag, New York, 1982).

[7] Brown, Kenneth S., 'The geometry of rewriting systems: A proof of the Anick-Groves-Squier Theorem' (preprint).

[8] Groves, J. R. J., and Smith, G. S., 'Rewriting systems and soluble groups' (preprint).

[9] Hilton, P. J., and Stammbach, U., *A course in homological algebra* (Springer-Verlag, New York, 1971).

[10] Le Chenadec, Ph., *Canonical forms in finitely presented algebras* (Pitman, London; Wiley, New York; 1986).

[11] Squier, Craig C., 'Word problems and a homological finiteness condition for modules', *J. Pure Appl. Algebra* **40** (1987), 201-217.

University of Melbourne
PARKVILLE 3052
Australia

TRANSVERSALS AND CONJUGACY IN THE GROUP OF RECURSIVE PERMUTATIONS

GRAHAM HIGMAN

1. INTRODUCTION

We are concerned in this paper with the relation of conjugacy in the group R of recursive permutations of the set N of natural numbers. As far as I know, the only previous work in this area is by Clement F. Kent, who proves in [2] the following theorem.

THEOREM 1.1 (Kent [2]). *If C is a conjugacy class of the full symmetric group S on N then the intersection $C \cap R$, assumed nonempty, is a single conjugacy class of R if and only if each element of C has only a finite number of infinite cycles.*

He further states in a footnote that at least one class of S contains an infinity of classes of R. Given the advances in recursion theory that have been made since Kent wrote, it is in fact rather easy to see that every class C of S whose elements have an infinity of infinite cycles and for which $C \cap R$ is nonempty contains an infinity of classes of R. Suppose, in fact, here and throughout the paper, that C_0 is the class of S consisting of permutations with an infinity of infinite cycles and no finite cycles. If C is any other class of S whose elements have an infinity of infinite cycles and $C \cap R$ is nonempty, then a short mainly group theoretic argument provides a natural injection of the set of classes of R in C_0 into the set of such classes in C. Thus it is sufficient to consider C_0. For an element ρ of C_0, a *transversal* of ρ is a subset of N which contains just one element from each cycle of ρ; in particular the *principal transversal* is that transversal which contains the smallest number in each cycle. If ρ is recursive it is easy to see that the principal transversal is Turing reducible to any transversal. But if T is a transversal of ρ, and σ is any permutation of N, $T\sigma$ is a transversal of $\sigma^{-1}\rho\sigma$, so that elements of C_0 which are conjugate by elements of R have, up to recursive equivalence, the same transversals. It follows that the Turing degree of the principal transversal of an element ρ of $C_0 \cap R$ is an invariant of its conjugacy class in R. A straightforward construction shows that this Turing degree can be any recursively enumerable degree, and since there are an infinity of these, we are home. (For recursion theoretic theorems quoted without proof, and recursion theoretic ideas introduced without definition, see Rogers [3]). These matters, and a few others of like triviality, are discussed in Section 2.

Section 3 is concerned with the possibility that an element ρ of $C_0 \cap R$ may not be conjugate in R to its nontrivial powers. This cannot be shown by the arguments of Section 2, because the principal transversals of ρ and of ρ^n, $n \neq 0$, are Turing equivalent. The easiest example, perhaps, is obtained by adapting the classical construction of a recursively enumerable set with immune complement to yield a permutation ρ in $C_0 \cap R$ all of whose transversals are immune, so that every infinite recursively enumerable subset of N contains two elements in the same cycle of ρ. Then an easy group theoretic argument, or an equally easy graph theoretic argument, shows that ρ is not conjugate to ρ^n for $|n| > 1$; but I do not know whether ρ may not be conjugate to ρ^{-1}, or ρ^2 to ρ^3. To remedy this defect we make another construction, ensuring nonconjugacy of powers directly. The result of this construction is an element ρ of $C_0 \cap R$ such that $\langle \rho \rangle$ is a retract of the approximate normaliser $N_R^*(\rho)$ (the set, that is, of elements σ of R such that $\sigma^{-1} \rho^m \sigma = \rho^n$ for some nonzero integers m and n). It follows not only that ρ^m is conjugate to ρ^n in R only if $m = n$, but also that $\langle \rho \rangle$ is a maximal cyclic subgroup of R. This example raises the question of how small $N_R^*(\rho)$ can be, for elements ρ of $C_0 \cap R$. Is it, for example, possible to have $N_R^*(\rho) = \langle \rho \rangle$? I would prefer not even to guess.

In Section 4, we raise the question whether the results of Section 2 could be improved by replacing the Turing degrees by degrees with respect to a stronger reducibility. We prove that they cannot be replaced by bounded-degrees, that is, degrees with respect to a reducibility that bears much the same relation to Turing reducibility that reducibility by bounded truth tables does to truth-table reducibility. We leave open the possibility that they could be replaced by truth-table degrees, though this seems unlikely. For recursion theoretic questions raised by this discussion, we refer the reader to Section 4 itself.

2. TRIVIALITIES

LEMMA 2.1. *If P is the principal transversal and T any transversal of the element ρ of $C_0 \cap R$ then P is Turing reducible to T.*

The function f which assigns to each natural number m the element of T contained in the cycle $m\langle \rho \rangle$ is recursive in T, because we can evaluate it by calculating the elements m, $m\rho$, $m\rho^{-1}$, $m\rho^2$, $m\rho^{-2}$, ... in turn till we find the one that is in T. But m belongs to P if and only if $f(m) \neq f(i)$ for $i = 0, 1, \ldots, m-1$. Thus $P \leqslant_T T$, as required.

THEOREM 2.2. *The Turing degree of the principal transversal of an element of $C_0 \cap R$ is an invariant of the conjugacy class of R containing it.*

Let P, Q be the principal transversals of elements ρ, σ of $C_0 \cap R$, and suppose that $\sigma = \kappa^{-1}\rho\kappa$ with κ recursive. Then $P\kappa$ is a transversal of σ, so that Lemma 2.1 gives $Q \leqslant_T P\kappa \equiv_T P$. Similarly $P \leqslant_T Q\kappa^{-1} \equiv_T Q$. Thus $P \equiv_T Q$, as required.

THEOREM 2.3. *The following three conditions are together necessary and sufficient for the subset P of N to be the principal transversal of an element of $C_0 \cap R$:*

(a) *both P and $\mathsf{N} \setminus P$ are infinite;*

(b) *$\mathsf{N} \setminus P$ is recursively enumerable;*

(c) *0 belongs to P.*

If P is the principal transversal of the element ρ of $C_0 \cap R$ and m belongs to $\mathsf{N} \setminus P$, this fact can be established by a finite amount of computation, for it means that $m > m\rho^i$ for some integer i. Thus $\mathsf{N} \setminus P$ is recursively enumerable, which establishes the necessity of (b). Since the necessity of (a) and of (c) is even more obvious, the burden of the proof is in establishing sufficiency.

Suppose then that P is a subset of N satisfying (a), (b) and (c). We shall construct a permutation ρ in $C_0 \cap R$ having P as its principal transversal by making definitions $m\rho = n$ one at a time. Because $\mathsf{N} \setminus P$ is infinite and recursively enumerable we can put $\mathsf{N} \setminus P = \{k_0, k_1, k_2, \ldots\}$ where k_i is a recursive function of i and the sequence has no repetitions. We are constructing an element of C_0, so we shall suppose that at no point when a finite number of definitions $m\rho = n$ have been made do we already have a finite cycle of ρ. Thus the definitions already made organise N into *maximal partial cycles* (MPC's), sequences $c_1, \ldots, c_s$ that is, such that definitions $c_i\rho = c_{i+1}$ have already been made for $i = 1, \ldots, s-1$, but we have made no definition $d\rho = c_1$ or $c_s\rho = d$. Of course, all but a finite number of MPC's consist of a single number c such that neither $c\rho$ nor $c\rho^{-1}$ has yet been defined. We shall furthermore suppose that at the point when r definitions $m\rho = n$ have been made the set $\{k_0, k_1, \ldots, k_{r-1}\}$ is precisely the set of numbers which are not the smallest in their MPC's. This is clearly true when $r = 0$, since the set is then empty, and every number is in an MPC by itself. So to define the construction, we may assume that it holds when r definitions have been made, for some fixed number r, and explain how the $(r+1)$-st definition is to be made, so that it continues to hold. Now $k_r \neq k_i$, if $i < r$, so that, by assumption, when r definitions have been made, k_r is the least element in the MPC containing it, which we denote by C. By (c) $k_r \neq 0$, so that there is another MPC, say C', containing a number less then k_r. We can make a definition $m\rho = n$ which fuses C and C' into a single MPC, either by taking m to be the last number in C and n the first in C', or by taking m to be last number in C' and n the first in C. In the fused MPC, k_r will not be the smallest element; but any number $j \neq k_r$ will be the smallest element in its MPC if and only if it was so before we made the definition. That is, the set of numbers which are not the smallest in their MPC's is now $\{k_0, k_1, \ldots, k_r\}$, as required.

There is a finite set of possibilities for C', so that we have a finite nonempty set of possibilities for the $(r+1)$-st definition $m\rho = n$. We choose between them by insisting first that $\min(m,n)$ is as small as possible; then, if necessary, that $\max(m,n)$ is as small as possible consistent with this; and finally, if there are still two choices left, that $m < n$. This completes the description of the construction; all that remains is to show that it works.

First, what emerges is a permutation; that is, for every number m both $m\rho$ and $m\rho^{-1}$ are sooner or later defined. Indeed, the first two definitions made are $0\rho = k_0$ and $k_1\rho = 0$; and if, for any number i, both $j\rho$ and $j\rho^{-1}$ have been defined for all $j < i$ by the time the r-th definition is made, then both $i\rho$ and $i\rho^{-1}$ will be defined by the time the s-th definition is made, if there are two numbers t in $r \leqslant t < s$ for which $k_t > i$, which will surely be true for large enough s, since $\mathsf{N} \setminus P$ is infinite. Thus ρ is a permutation, and clearly recursive. If ρ had a finite cycle this would show up at some finite point of the construction, which is not so. Moreover a number is not the smallest in the cycle of ρ containing it if and only if, at some finite point of the construction, it is not the smallest number in its MPC. By construction this is so if and only if the number belongs to $\mathsf{N} \setminus P$. Since P as well as $\mathsf{N} \setminus P$ is infinite, this implies that ρ belongs to C_0 and that P is its principal transversal. We are done.

COROLLARY 2.4. *Every recursively enumerable Turing degree is the degree of the principal transversal of an element of $C_0 \cap R$.*

COROLLARY 2.5. *$C_0 \cap R$ contains an infinity of conjugacy classes of R.*

We shall see later that in some cases the set of elements of $C_0 \cap R$ with principal transversal of given Turing degree consists of infinitely many conjugacy classes of R, and presumably this is what usually happens. There is, however, one case in which it does not. If π is the permutation of $\mathsf{N} \times \mathsf{Z}$ defined by $(m,n)\pi = (m, n+1)$, $m \in \mathsf{N}$, $n \in \mathsf{Z}$, then the elements $\theta^{-1}\pi\theta$ for recursive bijections $\theta \colon \mathsf{N} \times \mathsf{Z} \to \mathsf{N}$ form a conjugacy class of R contained in $C_0 \cap R$. We refer to it as the *standard class* and describe its elements as *standard*. The following theorem is essentially in [2].

THEOREM 2.6. *For an element ρ of $C_0 \cap R$ the following conditions are equivalent:*
  (i) *the principal transversal of ρ is recursive;*
  (ii) *ρ has a recursive transversal;*
  (iii) *ρ has a recursively enumerable transversal;*
  (iv) *ρ is standard.*

Trivially (i) $\Rightarrow$ (ii) $\Rightarrow$ (iii), and by Lemma 2.1 (ii) $\Rightarrow$ (i). If ρ has the recursively enumerable transversal T then there is a recursive function f which enumerates T without repetition, and θ defined by $(m,n)\theta = f(m)\rho^n$ is a recursive bijection $\mathsf{N} \times \mathsf{Z} \to \mathsf{N}$

with $\rho = \theta^{-1}\pi\theta$, so that ρ is standard. That is, (iii) $\Rightarrow$ (iv). But if $\theta\colon \mathbb{N} \times \mathbb{Z} \to \mathbb{N}$ is any recursive bijection the image under θ of $\{\,(m,0) \mid m \in \mathbb{N}\,\}$ is a recursive transversal of $\rho = \theta^{-1}\pi\theta$, so that (iv) $\Rightarrow$ (ii). The theorem follows.

Thus the standard class is characterised by the Turing degree of the principal transversal of any of its elements. That this is not so for all classes follows from the next theorem, together with either of the examples in Section 3.

THEOREM 2.7. *If ρ is an element of $\mathcal{C}_0 \cap R$ and n is a nonzero integer then the principal transversals of ρ and ρ^n are Turing equivalent.*

The principal transversals of ρ and ρ^{-1} are the same, so we may assume that $n > 0$. Let P, Q be the principal transversals of ρ and of ρ^n. Then

$$P = \{\, m \mid m \text{ is the smallest of a set of } n \text{ elements of } Q \text{ all in the same orbit of } \rho \,\}$$

and $\mathbb{N} \setminus P$ is recursively enumerable, so that $P \leqslant_T Q$. On the other hand

$$P^* = \{\, t\rho^i \mid t \in P,\ i = 0,1,\ldots,n-1 \,\}$$

is a transversal of ρ^n with $P^* \leqslant_T P$, so that by Lemma 2.1 $Q \leqslant_T P$. Hence $P \equiv_T Q$ as required.

We turn now to the connection between the decomposition of $\mathcal{C}_0 \cap R$ into conjugacy classes of R and the corresponding decomposition of $\mathcal{C} \cap R$, for any class of S each of whose elements has an infinity of infinite cycles. If ρ is any element of S we write $F(\rho)$ for the union of the finite cycles of ρ, and we consider first the case when, for elements ρ of $\mathcal{C}$, $F(\rho)$ is finite, say of size n. Put $I = \{0,1,\ldots,n-1\}$ and let R_I be the setwise stabiliser of I in R. Then R_I is isomorphic to $S_n \times R$, where S_n is the symmetric group on I, under an isomorphism which takes ρ in R_I to (α,λ), where α is the restriction of ρ to I, and λ is defined by $m\lambda = (m+n)\rho - n$, so that intuitively it is the restriction of ρ to $\mathbb{N} \setminus I$, "transferred to $\mathbb{N}$". Let $\mathcal{C}_f$ be the conjugacy class of S_n consisting of elements whose cycle structure is the finite part of the cycle structure of elements of $\mathcal{C}$. If α belongs to $\mathcal{C}_f$ and λ to $\mathcal{C}_0$, the element ρ of R_I which maps to (α,λ) belongs to $\mathcal{C}$; and if σ, mapping to (β,μ), is another such then, if λ is conjugate to μ in R, ρ is conjugate to σ in R_I, since α,β are conjugate in S_n, and hence ρ is conjugate to σ in R. That is, mapping the conjugacy class of R containing λ to the conjugacy class containing ρ is a well defined map from the set of conjugacy classes of R in $\mathcal{C}_0 \cap R$ to the set in $\mathcal{C} \cap R$. This map is an injection, because if ρ, σ are conjugate in R the conjugating element must lie in R_I, since $I = F(\rho) = F(\sigma)$, whence λ, μ are conjugate in R. It is surjective, because R is transitive on the n-element subsets of $\mathbb{N}$, so that every conjugacy class of R in $\mathcal{C} \cap R$ contains an element ρ with $F(\rho) = I$. That is, we have a natural bijection between the set of conjugacy classes of R in $\mathcal{C}_0 \cap R$ and the set in $\mathcal{C} \cap R$.

If we try to make a similar argument for a class C such that $F(\rho)$ is infinite for ρ in C, we strike two snags. The first is that $C \cap R$ may well be empty. Of course we can confine attention to cases where it is not, but then we shall need the following lemma, where now C_f denotes the conjugacy class of S whose elements have the same cycle structure as the restriction of ρ to $F(\rho)$, where ρ belongs to C.

LEMMA 2.8. If $C \cap R$ is nonempty then so is $C_f \cap R$.

If ρ belongs to $C \cap R$ then $F(\rho)$ is recursively enumerable and infinite, so that there is a recursive function f that enumerates it without repetitions. The equation $f(n)\rho = f(n\sigma)$ then defines an element σ of $C_f \cap R$, which proves the lemma.

Notice that it follows from Theorem 1.1 that if $C_f \cap R$ is nonempty then it is a single conjugacy class of R. Armed with this we construct in the same way as before an injection from the set of conjugacy classes of R in $C_0 \cap R$ to the set in $C \cap R$, merely replacing I by a fixed infinite coinfinite recursive set, say the set E of even integers. (Thus R_E is isomorphic to $R \times R$). We strike the second snag if we try to prove that this injection is surjective, because in general it is not. Its image is the set of classes in $C \cap R$ containing elements ρ such that $F(\rho)$ is recursive. This is the set of all classes in $C \cap R$ if, for an element ρ of C, there is a bound to the lengths of the finite cycles of ρ, for then $F(\rho)$ is the fixed point set of ρ^n, where n is the least common multiple of these lengths. Otherwise it is not, though we refrain from producing the examples which prove this. We sum up:

THEOREM 2.9. If C is a conjugacy class of S each of whose elements has an infinity of infinite cycles then either $C \cap R$ is empty or there is a natural injection from the set of conjugacy classes of R in $C_0 \cap R$ to the set in $C \cap R$, which is a bijection if there is a bound to the lengths of the finite cycles of elements of C. In particular, if $C \cap R$ is nonempty it contains an infinity of classes of R.

3. NONCONJUGACY OF POWERS

Recall that a subset of $\mathbb{N}$ is said to be *immune* if it is infinite but has no infinite recursively enumerable subset. Thus all transversals of the element ρ of $C_0 \cap R$ are immune if and only if every infinite recursively enumerable set contains two elements in the same cycle of ρ.

THEOREM 3.1. There exists an element of $C_0 \cap R$ all of whose transversals are immune.

We construct such a permutation ρ by making definitions $m\rho = n$, or equivalently $n\rho^{-1} = m$, one at a time, avoiding the creation of finite cycles of ρ, so that, as in the proof of Theorem 2.3, the finite set of definitions that has been made at any point organises $\mathbb{N}$ into MPC's. There exists a sequence X_0, X_1, ... of subsets of $\mathbb{N}$ which includes each recursively enumerable set at least once, such that the set of true statements 'j belongs to X_i' is recursively enumerable. Our description of the way that ρ is constructed is in terms of the output of a machine M which enumerates these statements. Whenever M prints a statement 'j belongs to X_i' we do two things. First we check through the sequence 0ρ, $0\rho^{-1}$, 1ρ, $1\rho^{-1}$, 2ρ, ... to find the first term $m\rho^\epsilon$, $\epsilon = \pm 1$ that has not already been defined, and make the definition $m\rho^\epsilon = r$, where r is the smallest number greater than $4m$ such that neither $r\rho$ nor $r\rho^{-1}$ has yet been defined. Secondly we check whether it is necessary and possible, under rules to be described, to act to ensure that X_i contains two numbers in the same cycle of ρ. It is necessary so to act unless there are two numbers k_1 and k_2 in the same MPC such that M has already made the statements 'k_1 belongs to X_i' and 'k_2 belongs to X_i'. It is possible so to act if M has already made such statements, but k_1 and k_2 belong to different MPC's, at least one of which contains no number less then $4i$. A possible action is to make a definition $m\rho = n$ which fuses these two MPC's. If it is either unnecessary or impossible to take action we do nothing. But if it is both necessary and possible, we choose in some systematic way between the finite number of possible actions. (Notice that then it will never again be necessary to take action in respect of the same X_i). Since M has an infinity of statements to make the construction proceeds indefinitely. What emerges is clearly a recursive permutation ρ without finite cycles. Moreover, for any natural number k, the number of MPC's containing numbers less then $4k$ can diminish only when we make a definition to ensure that $m\rho^\epsilon$ is defined, for $m < k$, or when we act to ensure that two elements of X_i belong to same cycle of ρ, for $i < k$; hence on at most $3k$ occasions. When it does diminish, it diminishes by 1. Thus the number of cycles of ρ that contain numbers less than $4k$ is at least k, so that ρ has an infinity of cycles and so belongs to $\mathcal{C}_0$. Finally if X_i is infinite then at some point M makes its $(4i+1)$-st statement 'j belongs to X_i' (or its 2nd, if $i = 0$). At this point, if it is necessary to act it is also possible; for if the $4i+1$ numbers known to belong to X_i are in different MPC's, one of these MPC's must contain no number less than $4i$. Thus X_i contains two numbers in the same cycle of ρ, which completes the proof of the theorem.

THEOREM 3.2. *If ρ is an element of $\mathcal{C}_0 \cap R$ all of whose transversals are immune then ρ is not conjugate in R to ρ^n for any integer n for which $|n| > 1$.*

If ρ, σ are elements of R with $\sigma^{-1}\rho\sigma = \rho^n$, where $|n| > 1$, then for any natural number m the set $X = \{ m\sigma^i\rho\sigma^{-i}, i = 0,1,2,\ldots \}$ is recursively enumerable. Suppose that $m\sigma^i\rho\sigma^{-i} = m\sigma^j\rho\sigma^{-j}\rho^r$ for integers i, j, r with $j > i \geqslant 0$. Then a short calculation shows that $m\sigma^j$ is a fixed point of $\rho^{rn^j - n^{j-i}+1}$; and since $|n| > 1$, $rn^j - n^{j-i}+1 \neq 0$. Thus either ρ has a finite cycle or the elements of X are distinct as written and belong to different cycles of ρ. That is, either ρ does not belong to C_0 or it has an infinite recursively enumerable partial transversal, which implies the theorem.

REMARK. There is a graph theoretic proof of Theorem 3.2 which conceptually is at least as simple as the proof given above. If ρ, σ are elements of S such that $\sigma^{-1}\rho\sigma = \rho^n$ and ρ belongs to C_0 then for each cycle C of ρ, $C\sigma$ is a cycle of ρ^n, and, of course, each cycle of ρ is the union of $|n|$ cycles of ρ^n. We form a directed graph Γ by taking as vertices the cycles of ρ, with an edge from cycle C_1 to C_2 if and only if $C_1\sigma \subset C_2$. Then at each vertex of Γ there is just one outward edge and just $|n|$ inward edges. Connected graphs of this form are easy to classify: there is one in which the underlying undirected graph is a tree of valency $|n| + 1$; and for each positive integer k there is one which has a circuit of length k at every vertex of which hangs a tree. In all cases there are many integers m such that m, $m\sigma^{-1}$, $m\sigma^{-2}$, ... belong to different cycles of ρ, which proves the theorem.

LEMMA 3.3. *Let E be a recursively enumerable equivalence relation on $\mathbb{N}$. If there exists an infinite recursively enumerable set X and an integer n such that X meets each equivalence class of E in at most n elements, then there exists an infinite recursively enumerable set Y all of whose elements lie in different equivalence classes of E.*

Let m be the largest number such that there is an infinity of equivalence classes of E each containing exactly m elements of X. By throwing away a finite set of numbers from X, if necessary, we may assume that no equivalence class of E contains more than m elements of X. Then

$$Y = \big\{\, k \mid k \text{ is the smallest of a set of } m \text{ elements of } X \text{ all equivalent under } E \,\big\}$$

has the desired property.

THEOREM 3.4. *If the element ρ of $C_0 \cap R$ has all its transversals immune, so does ρ^n, for any $n \neq 0$.*

We may assume that $n > 0$ and then the theorem follows immediately from Lemma 3.3, taking E to be the relation of belonging to the same cycle of ρ.

Taking Theorem 3.2 and Theorem 3.4 together gives immediately:

THEOREM 3.5. *If ρ is an element of $C_0 \cap R$ with all its transversals immune, and n_0, n_1, n_2, ... are nonzero integers such that, for all i, n_{i+1} is a proper multiple of n_i, then ρ^{n_i} and ρ^{n_j} are conjugate in R only if $i = j$.*

Theorem 3.5 of course leaves open the question whether, for instance, under its hypotheses ρ^2 is nonconjugate to ρ^3. Neither of the suggested proofs of Theorem 3.2 is very helpful here. The group theoretic proof depends on the simple structure of the group $\langle \rho, \sigma \, ; \, \sigma^{-1} \rho \sigma = \rho^n \rangle$, which is not shared by the group $\langle \rho, \sigma \, ; \, \sigma^{-1} \rho^2 \sigma = \rho^3 \rangle$; and an analogue of the graph theoretic proof sketched in the Remark requires an analysis of directed graphs with an outward valency of 2 and an inward valency of 3, which is probably difficult. So to prove the strongest nonconjugacy property of powers, we make a new construction. But first we place our considerations in context.

Recall (for instance from [1]) that if g is an element of infinite order in a group G then the *approximate centraliser* $C_G^*(g)$ is the union of the centralisers of the nontrivial powers of g, and the *approximate normaliser* $N_G^*(g)$ the set of elements t of G such that $t^{-1} g^m t = g^n$ for some nonzero integers m, n. Both $C_G^*(g)$ and $N_G^*(g)$ are groups. For an element ρ of $\mathcal{C}_0$, fairly complete descriptions of $C_S^*(\rho)$ and $N_S^*(\rho)$ are given in [1]. We summarize them, insofar as they are illuminating here.

First, $C_S^*(\rho)$ is normal in $N_S^*(\rho)$ and the factor group is isomorphic to the multiplicative group $\mathbf{Q}^*$ of nonzero rationals. Next $C_S^*(\rho)$ has a chain $1 < K < L < M$ of characteristic subgroups. Here K is a countable locally abelian-by-finite simple group, and $C_S^*(\rho)/M$ is an unmanageable uncountable simple group: one, for instance, that embeds every countable group. M/K is torsionfree abelian, L/K is its maximum divisible subgroup, and is uncountable, and M/L is isomorphic to the profinite completion $\hat{\mathbf{Z}}$ of the additive group $\mathbf{Z}$. Lastly $C_S^*(\rho)$ has a subgroup N such that $K < N < M$, not obviously characteristic but nevertheless normal in $N_S^*(\rho)$, such that N/K is infinite cyclic, and L/K and N/K generate their direct product LN/K, LN being the preimage of $\mathbf{Z}$ in the natural map of M to $\hat{\mathbf{Z}}$ with kernel L. Of these groups, N is the easiest to describe: it is the set of those elements of $N_S^*(\rho)$ that fix every number outside some finite set of cycles of ρ. In what follows we shall write $K(\rho)$, $L(\rho)$ etc. for K, L etc., to emphasize the dependence of these groups on ρ. Note that $N(\rho)$ is *not* a normaliser.

Of course, if ρ belongs to $\mathcal{C}_0 \cap R$, $C_R^*(\rho) = C_S^*(\rho) \cap R$ and $N_R^*(\rho) = N_S^*(\rho) \cap R$. It would be straightforward to rework the material in [1] to show that if σ is a standard element of $\mathcal{C}_0 \cap R$ then $N_R^*(\sigma)$ has a structure rather similar to that described above for $N_S^*(\rho)$. The unrestricted permutational wreath product $\mathbf{Z} \operatorname{Wr} S$, that plays a large part in [1], would have to be replaced by its constructive analogue $B_{\mathrm{rec}} R$, where B_{rec} is the additive group of recursive functions $\mathbf{N} \to \mathbf{Z}$; and the result of Kent [2] that S_0 and A (the group of permutations with finite support and the alternating group) are the only proper nontrivial normal subgroups of R would be used instead of the corresponding fact for S. The resulting argument would show that $N_R^*(\sigma)/C_R^*(\sigma)$ is isomorphic to $\mathbf{Q}^*$, and that $C_R^*(\sigma)/(M(\sigma) \cap R)$ is a rather unmanageable countable simple group; one, for instance, that embeds every finitely generated group with solvable word problem. $N(\sigma)$, and hence also $K(\sigma)$, is contained in R, and the argument would further show that $(L(\sigma) \cap R)/K(\sigma)$ is a torsionfree divisible abelian group of countably infinite rank, and that $(M(\sigma) \cap R)/(L(\sigma) \cap R)$ is isomorphic to the subgroup of $\widehat{Z}$ consisting of limits of recursive sequences of integers.

All this amounts to saying that if σ is standard then $N_R^*(\sigma)/N(\sigma)$ is a large and complicated group. What we construct below, in contrast, is an element ρ of $C_0 \cap R$ such that $N_R^*(\rho) = \langle \rho \rangle N(\rho) \cap R$. As far as I can see the method gives no information about $N(\rho) \cap R$, and any attempt to improve on the result would have to do something about this. We state the theorem in the terms which come most naturally out of the construction.

THEOREM 3.6. *There exists an element ρ of $C_0 \cap R$ such that for any element π of $N_R^*(\rho)$ there exists an integer n and a finite union X of cycles of ρ such that for all m in $\mathbb{N} \backslash X$ we have $m\pi = m\rho^n$.*

A recursive partial permutation of $\mathbb{N}$ is a bijection π between subsets of $\mathbb{N}$ such that the set of pairs $(a, a\pi)$ is recursively enumerable. It is, of course, possible to construct a sequence π_0, π_1, π_2, ... of recursive partial permutations which contains each recursive partial permutation at least once such that the set of true statements '$a\pi_i = b$' is recursively enumerable. Our construction will be based on the output of a machine M which enumerates this set of statements. We shall make definitions $m\rho = n$ one at a time, or at least in finite blocks, avoiding definitions which create a finite cycle, and so organising $\mathbb{N}$, at each point in the construction, into MPC's. Whenever the machine M makes a new statement, we do two things. First we check through the sequence 0ρ, $0\rho^{-1}$, 1ρ, $1\rho^{-1}$, 2ρ, ... to find the first term $m\rho^\epsilon$ that has not yet been defined, and put $m\rho^\epsilon = r$, where r is the smallest number greater than 2^m such that neither $r\rho$ nor $r\rho^{-1}$ has yet been defined. Secondly we attend, according to a procedure to be described, to a requirement (p, q, i). There is one of these requirements for each triple (p, q, i) where p is a positive integer, q a nonzero integer and i a natural number: attending to the requirement is designed to ensure, if possible, that π_i is not a permutation such that $\pi_i^{-1} \rho^p \pi_i = \rho^q$. The order in which the requirements are given attention is settled in advance, by a recursive rule that ensures that each is given attention infinitely often.

To attend to the requirement (p, q, i) we first ask whether there are numbers a, b, c, d such that M has already made the statements '$a\pi_i = c$' and '$b\pi_i = d$', and definitions $m\rho = n$ have been made which imply that $a\rho^{kp} = b$ and $c\rho^r = d$, where k, r are integers such that $r \neq kq$. If so action is *unnecessary* and we do nothing. Otherwise we go on to ask whether action is *possible*. Action is possible if, for numbers a, b, c, d, M has already made the statements '$a\pi_i = c$' and '$b\pi_i = d$' and either (i) a, b, c, d belong to distinct MPC's $C(a), C(b), C(c)$ and $C(d)$, none of which contains a number less than $2^{\max(p, |q|, i)}$ or (ii) definitions $m\rho = n$ already made imply that $a\rho^s = c$, $b\rho^t = d$ with $s \neq t$, but a, b are in distinct MPC's $C(a)$ and $C(b)$, neither of which contains a number less than $2^{\max(p, |q|, i)}$. The action to be taken, in both cases, is to fuse the MPC's $C(a)$ and $C(b)$, possibly with padding, that is to make definitions $e_0\rho = e_1$, $e_1\rho = e_2$, ..., $e_j\rho = e_{j+1}$ for some $j \geqslant 0$, where e_0 is the last element of $C(a)$ and e_{j+1} the first element of $C(b)$; and in case (i) also to fuse $C(c)$ and $C(d)$ without padding. The amount of padding is to be adjusted so that $a\rho^{kp} = b$ and $c\rho^r = d$ for integers k, r such that $r \neq kq$. It is obvious that this can be done in case (i); in case (ii) we can make $a\rho^{kp} = b$ for any sufficiently large k, and then $c\rho^r = d$ with $r = kp - s + t$, so that we only have to avoid making $k(p - q) = s - t$, which is possible since $s \neq t$. If action is both necessary and possible we take it, choosing in some systematic way between the possible sets $\{a, b, c, d\}$, using as little padding as possible, and taking the padding elements to be the smallest available numbers greater than $2^{\max(p,|q|,i)}$.

This concludes the description of the way we attend to the requirement (p, q, i) and hence of the way we construct ρ; but before going on to check that the construction does what is required of it we make two observations. First, in taking action to attend to the requirement (p, q, i) we make at most $2p$ definitions. Thus for any number k the number of MPC's containing numbers less than k is diminished by at most $2p$. If $k \leqslant 2^{\max(p,|q|,i)}$ the number is unaltered. Furthermore once such action is taken, it becomes unnecessary ever to take it again. Secondly, if, at any point in the construction which calls for attention to the requirement (p, q, i), action is possible then either it is unnecessary or we take it. In either case we end up with numbers a, b, c, d such that $a\pi_i = c$, $b\pi_i = d$, $a\rho^{kp} = b$, $c\rho^r = d$ where $r \neq kq$, and since ρ has no finite cycles this implies that π_i is not a permutation such that $\pi_i^{-1}\rho^p\pi_i = \rho^q$.

It is evident that the construction yields a permutation ρ with no finite cycles. The number of MPC's containing numbers less than 2^k for any natural number k is diminished during the construction only when we define $m\rho^\epsilon$ for some $m < k$ and $\epsilon = \pm 1$, when it may diminish by 1, and when we take action in attending to a requirement (p, q, i) with $p < k$, $|q| < k$ and $i < k$, when it diminishes by at most $2p$. Thus the number of cycles of ρ containing numbers less than 2^k is at least $2^k - 4k^4 - 2k$, and so tends to infinity with k. That is, ρ has an infinity of cycles, and so belongs to $\mathcal{C}_0$. Now suppose that π belongs to $N_R^*(\rho)$, so that $\pi^{-1}\rho^p\pi = \rho^q$ for some nonzero integers p, q, where we may suppose p positive. Then $\pi = \pi_i$ for some i. Suppose first that there is no finite union X of cycles of ρ such that, for all m in $\mathbb{N} \setminus X$, $m\pi_i$ belongs to the same cycle of ρ as m. Let X_0 be the union of the cycles of ρ that contain numbers less than $2^{\max(p,|q|,i)}$. Because $\pi_i^{-1}\rho^p\pi_i = \rho^q$, π_i maps each cycle of ρ^p onto a cycle of ρ^q, and hence there is a finite union X_1 of cycles of ρ which contains $X_0\pi_i^{-1}$. By assumption there is a number a not in $X_0 \cup X_1$ such that a and $c = a\pi_i$ belong to different cycles of ρ, and by construction neither of these cycles contains a number less than $2^{\max(p,|q|,i)}$. Repeating the argument with X_0 replaced by $X_0 \cup a\langle\rho\rangle \cup c\langle\rho\rangle$ we obtain two further numbers b, d such that $b\pi_i = d$ and a, b, c, d belong to different cycles of ρ none of which contains a number less than $2^{\max(p,|q|,i)}$. But then at any point in the construction after M has made the statements '$a\pi_i = c$' and '$b\pi_i = d$' at which attention to the requirement (p,q,i) is called for, action is possible. But then π_i cannot be a permutation such that $\pi_i^{-1}\rho^p\pi_i = \rho^q$. Hence there exists a finite union X of cycles of ρ such that, for m in $\mathsf{N}\setminus X$, $m\pi_i = m\rho^n$ where n may depend on m. If X cannot be chosen so that n is independent of m, an argument similar to the one above produces numbers a, b in different cycles of ρ neither of which contains a number less then $2^{\max(p,|q|,i)}$ such that $c = a\pi_i = a\rho^s$ and $d = b\pi_i = b\rho^t$ with $s \neq t$, leading to the same contradiction. That is, there is a finite union X of cycles of ρ and an integer n such that, for m in $\mathsf{N}\setminus X$, $m\pi = m\pi_i = m\rho^n$. Since π was any element of $N_R^*(\rho)$, we are done.

THEOREM 3.7. *If the element ρ of $C_0 \cap R$ satisfies the condition required of it in Theorem 3.6 then*

  (i) *$\langle\rho\rangle$ is a retract of $N_R^*(\rho)$;*
 (ii) *powers ρ^m and ρ^n of ρ are conjugate in R only if $m = n$;*
(iii) *$\langle\rho\rangle$ is a maximal cyclic subgroup of R; and*
 (iv) *$N_R^*(\rho) = \langle\rho\rangle N(\rho) \cap R$.*

(i) By assumption if π belongs to $N_R^*(\rho)$ there is a finite union X of cycles of ρ and an integer n such that, for m in $\mathsf{N}\setminus X$, $m\pi = m\rho^n$. The map carrying π to ρ^n is evidently an endomorphism $N_R^*(\rho) \to \langle\rho\rangle$ which is the identity on its image $\langle\rho\rangle$, as required. In what follows we denote the retraction by a bar.

(ii) If m,n are integers and π is an element of R such that $\pi^{-1}\rho^m\pi = \rho^n$ then if one of m,n is zero both are. Otherwise π is an element of $N_R^*(\rho)$ so we may apply the retraction to give $\bar\pi^{-1}\rho^m\bar\pi = \rho^n$. Since $\bar\pi$ is a power of ρ, $m = n$ as required.

(iii) If $\langle\pi\rangle$ is a cyclic subgroup of R containing $\langle\rho\rangle$ then $\rho = \pi^k$ for some integer k. Then π belongs to $N_R^*(\rho)$ so that we can apply the retraction to give $\rho = \bar\pi^k$. Here $\bar\pi$ is a power of ρ, say $\bar\pi = \rho^l$. Then $\rho = \rho^{kl}$, so that $k = l = \pm 1$, and $\langle\rho\rangle = \langle\pi\rangle$, as required.

(iv) If π belongs to the kernel of the retraction then $m\pi = m$ for m outside some finite union of cycles of ρ, so that π belongs to $N(\rho)$. Thus

$$N_R^*(\rho) = \langle\rho\rangle(N(\rho)\cap R) = \langle\rho\rangle N(\rho)\cap R,$$

which proves (iv).

4. STRONGER REDUCIBILITIES

In this section we consider the possibility that Theorem 2.2 could be strengthened by replacing the Turing degree by a degree with respect to a stronger reducibility. We construct an example which sets limits to what can be done in this direction. But our results are very incomplete, and we end with a brief discussion of this.

To explain exactly what the example achieves we recall some definitions and make some new ones. For subsets X, Y of N, to say that $X \leqslant_T Y$ means that there exist recursively enumerable sets U, V of triples (m, A, B) with m in N, A, B finite subsets of N, such that (i) m belongs to X if and only if, for some (m, A, B) in U, $A \subset Y$ and $B \subset \mathsf{N} \setminus Y$, and (ii) m belongs to $\mathsf{N} \setminus X$ if and only if, for some (m, A, B) in V, $A \subset Y$ and $B \subset \mathsf{N} \setminus Y$. More fully, one can then say that X *is Turing reducible to Y via* (U, V). We shall say that the set U of triples (m, A, B) is *bounded* if there is a number n such that for all (m, A, B) in U both $|A| \leqslant n$ and $|B| \leqslant n$. Accordingly, X *is boundedly reducible to Y*, written $X \leqslant_b Y$, if X is Turing reducible to Y via (U, V) where both U and V are bounded. The relation $\leqslant_b$ is transitive, and so gives rise to an equivalence relation: X *is boundedly equivalent to Y*, written $X \equiv_b Y$, if both $X \leqslant_b Y$ and $Y \leqslant_b X$. We call the classes of this equivalence relation *bounded-degrees*. To put these notions into context, observe that $X \leqslant_{btt} Y$ (bounded truth table reducibility) implies $X \leqslant_b Y$ and $X \leqslant_b Y$ implies $X \leqslant_T Y$, but that neither of these implications is obviously reversible, and that neither of $X \leqslant_{tt} Y$ and $X \leqslant_b Y$ obviously implies the other.

What we construct below is an element ρ of $\mathcal{C}_0 \cap R$ and an element κ of R such that for any two distinct integers p and q the principal transversals of $\kappa^{-p} \rho \kappa^p$ and $\kappa^{-q} \rho \kappa^q$ have incomparable bounded-degrees. Thus the Turing degree in Theorem 2.2 cannot be replaced by a bounded-degree, or *a fortiori* by a bounded truth-table degree, or by a degree with respect to any stronger reducibility. The example leaves open the possibility that it could be replaced by a truth-table degree, though this seems unlikely. The example of course shows incidentally that $X \leqslant_T Y$ does not imply $X \leqslant_b Y$.

It is clear that we can introduce in an entirely analogous manner a notion of bounded enumeration reducibility. For $X \leqslant_e Y$ if and only if $X \leqslant_e Y$ via some recursively enumerable set U of pairs (m, A), m in N, A a finite subset of N. We say that U is *bounded* if there is an integer n such that $|A| \leqslant n$ whenever (m, A) belongs to U, and that X is *boundedly enumeration reducible to Y*, written $X \leqslant_{be} Y$, if $X \leqslant_e Y$ via U for some bounded U. The point of this is that it is well known (and set as excercise 9–60 in [3]) that for sets P, Q whose complements are recursively enumerable $P \leqslant_T Q$ and $P \leqslant_e Q$ are equivalent; and the argument which proves this proves also that for such sets $P \leqslant_b Q$ and $P \leqslant_{be} Q$ are equivalent. Thus in Theorem 2.2 we could have written 'enumeration degree' instead of 'Turing degree', and in Theorem 4.1 below, 'bounded enumeration degrees' instead of 'bounded-degrees'. In the context of Theorem 2.2 Turing degrees seem a lot more natural, and since the sole interest of Theorem 4.1 is in the comparison with Theorem 2.2, this more or less forces the choice there also. But when we actually carry out the construction the gain in convenience from the simpler definition of enumeration reducibility is decisive, and we use that concept.

THEOREM 4.1. *There exist elements ρ of $C_0 \cap R$ and κ of R such that, for any two distinct integers p and q, the bounded-degrees of the principal transversals of $\kappa^{-p}\rho\kappa^p$ and $\kappa^{-q}\rho\kappa^q$ are incomparable.*

The construction of ρ is a fairly straightforward finite damage priority argument; but first we define the conjugating element κ. We choose a recursive sequence b_0, b_1, b_2, ... of positive integers which tends to infinity faster than 2^{i+1}, and define a_0, a_1, a_2, ... by $a_0 = 0$, $a_{i+1} = a_i + 1 + (i+1)b_i$. We then set $a_i\kappa = a_i$ for all i; and if $a_i < c < a_{i+1}$, we set $c\kappa = c + b_i$ unless $c + b_i \geqslant a_{i+1}$, in which case we set $c\kappa = a_i + a_{i+1} - c$. This means that if we divide the interval $\{\, c \mid a_i < c < a_{i+1} \,\}$, as we can, into $i + 1$ equal subintervals I_0, I_1, ..., I_i then the restriction of κ to I_α is the order preserving map from I_α to $I_{\alpha+1}$ for $\alpha = 0, \ldots, i-1$, and for $\alpha = i$ it is the order reversing map from I_i to I_0.

The construction of ρ uses the output of a machine M which enumerates the true statements '(k, A) belongs to U_j' where U_0, U_1, U_2, ... is a sequence of recursively enumerable sets of pairs (k, A), k in N, A a finite subset of N, which includes every such set at least once. We construct ρ by making definitions $m\rho = n$ one at a time, avoiding definitions which create a finite cycle of ρ, so that at any point in the construction the definitions already made organise N into MPC's for ρ. Of course, $m\rho = n$ implies $m\kappa^p\kappa^{-p}\rho\kappa^p = n\kappa^p$, for any integer p, so that the image under κ^p of an MPC for ρ is an MPC for $\kappa^{-p}\rho\kappa^p$. We denote by P_p the principal transversal of $\kappa^{-p}\rho\kappa^p$, so that at any point in the construction the numbers already known not to be in P_p are those which are not the smallest in their MPC's for $\kappa^{-p}\rho\kappa^p$. When we make a definition $m\rho = n$ we fuse two MPC's for ρ, say C_1 and C_2. If k_i, $i = 1, 2$, is the smallest number in C_i we now know, which we did not before, that $\max(k_1, k_2)$ is not in P_0. We say that making the definition *adds* $\max(k_1, k_2)$ *to* $\mathsf{N} \setminus P_0$. Similarly making the definition adds one number to $\mathsf{N} \setminus P_p$ for each integer p. Notice that since κ fixes each a_i and fixes setwise each interval $\{\, c \mid a_i < c < a_{i+1} \,\}$, if a_i is the smallest number in an MPC for ρ it is also the smallest number in the corresponding MPC for $\kappa^{-p}\rho\kappa^p$, and if the smallest number in an MPC for ρ lies in the interval $\{\, c \mid a_i < c < a_{i+1} \,\}$ so does the smallest number in the corresponding MPC for $\kappa^{-p}\rho\kappa^p$. It follows that when we make a definition $m\rho = n$ the number added to $\mathsf{N} \setminus P_p$ is either the same a_i for all p or lies in the same interval $\{\, c \mid a_i < c < a_{i+1} \,\}$ for all p.

At any point in the construction we shall have in front of us a list of the statements '(k, A) belongs to U_j' already made by M, and a list of the definitions $m\rho = n$ already made. From the second list we can determine, for any integer p, what numbers are already known to belong to $\mathsf{N} \setminus P_p$. If p, q are distinct integers and j is a natural number then by a *potential witness (PW)* against (p, q, j) we mean a pair (k, A) such that M has already made the statement '(k, A) belongs to U_j' and we already know that k belongs to $\mathsf{N} \setminus P_p$, but not that any element of A belongs to $\mathsf{N} \setminus P_q$. If this situation persists to the end of the construction (k, A) will become a witness to the fact that P_p is not enumeration reducible to P_q via U_j. But it may not persist, because a later definition $m\rho = n$ may add an element of A to $\mathsf{N} \setminus P_q$. If this happens we say that the definition *destroys* the PW (k, A). From time to time a PW against (p, q, j) may be elected *preferred potential witness (PPW)*; in addition to the lists already mentioned we keep lists of the current PPW's. The idea of the construction is, of course, that PPW's should not be destroyed. But because of conflicting requirements we cannot make this an absolute rule. We have to decide priorities between the triples (p, q, j). That is, we choose a recursive bijection f from the set of triples (p, q, j) to N, and say that (p, q, j) *has higher priority than* (p', q', j') if $f(p, q, j) < f(p', q', j')$.

The procedure for constructing ρ is as follows. Whenever the machine M makes a new statement we do two things. First we check through the sequence 0ρ, $0\rho^{-1}$, 1ρ, $1\rho^{-1}$, 2ρ, ... to find the first term $m\rho^\epsilon$ that has not yet been defined, and set $m\rho^\epsilon = a_i$, where i is as small as possible subject to the conditions that neither $a_i\rho$ nor $a_i\rho^{-1}$ has yet been defined, $m < a_i$, and the definition destroys no current PPW. (Notice that this definition adds a_i to $\mathsf{N} \setminus P_p$ for each integer p). Secondly we ensure, if possible, under rules to be stated, that a PPW exists against a certain triple (p, q, j). The particular triple to be considered is decided by a recursive rule, settled in advance, which guarantees that each triple is considered infinitely often. If there is already a PPW against (p, q, j) we do nothing. If there is no PPW but there are PW's we choose one of them in some systematic way and elect it PPW. If there are no PW's against (p, q, j) we consider the possibility of creating one and then electing it PPW. For this to be possible there must be a pair (k, A) such that M has already made the statement '(k, A) belongs to U_j' and we do not know either that k belongs to $\mathsf{N} \setminus P_p$ or that any element of A belongs to $\mathsf{N} \setminus P_q$. Then unless $k = 0$ there will be definitions $m\rho = n$ which, if we make them, will add k to $\mathsf{N} \setminus P_p$. Each such definition will add a number l to $\mathsf{N} \setminus P_q$, and if l belongs to A we shall be no better off. But if not, making the definition will turn (k, A) into a PW against (p, q, j) which can be elected PPW. We shall not do this unless two further conditions (called the *side conditions*) are met, namely that the number k added to P_p is not less than a_{i+1}, where $i = f(p, q, j)$, and that the definition destroys no PPW against a triple with higher priority than (p, q, j). If the side conditions can be met we choose in some systematic way between the definitions $m\rho = n$ which satisfy them, and create and elect a PPW against (p, q, j). This action may destroy PPW's against triples of lower priority than (p, q, j) and if so they must be removed from the list.

Since the machine M has an infinity of statements to make this process continues indefinitely. Evidently it yields a recursive permutation ρ with no finite cycles. We arrange the proof that it has the properties required of it in a series of lemmas.

LEMMA 4.2. *If* $f(p, q, j) = i$ *then the number of occasions throughout the construction on which a PPW is elected against a triple with a higher priority than* (p, q, j) *is finite, and indeed is less than* 2^i.

This is obvious if $i = 0$, so we use induction. Suppose that the number in question is finite and equal to n_i. Between any two occasions on which a PPW is elected against (p, q, j) the earlier one must be destroyed. This can happen only when a PPW against a triple with a higher priority is created and elected, and so on at most n_i occasions. Thus a PPW is elected against (p, q, j) on at most $n_i + 1$ occasions, giving $n_{i+1} \leqslant 2n_i + 1$, which provides the induction step.

LEMMA 4.3. *The number of elements of the interval* $\{ c \mid a_i < c < a_{i+1} \}$ *added to* $\mathsf{N} \setminus P_0$ *at any point in the construction is less than* 2^i.

If $f(p, q, j) = i$ a number in $\{ c \mid a_i < c < a_{i+1} \}$ can be added to $\mathsf{N} \setminus P_0$ only when we create and elect a PPW against a triple with a higher priority than (p, q, j); so that Lemma 4.3 follows from Lemma 4.2.

LEMMA 4.4. *The permutation* ρ *has an infinity of cycles and so belongs to* $C_0 \cap R$.

By construction the interval $\{ c \mid a_i < c < a_{i+1} \}$ contains more than 2^i elements, at least for large i. So each such interval, except perhaps for a finite number, contains an element of P_0 by Lemma 4.3. These elements lie in different cycles of ρ, which proves Lemma 4.4.

LEMMA 4.5. *Let* p, q *be distinct integers and put* $q - p = m$. *For* $i \geqslant |m|$ *there exist* $d = d(i)$ *numbers* $c_1, \ldots, c_d$ *in* $\{ c \mid a_i < c < a_{i+1} \}$ *such that*
  (i) $c_1 < c_2 < \cdots < c_d$;
  (ii) $c_1 \kappa^m > c_2 \kappa^m > \cdots > c_d \kappa^m$;
  (iii) $c_1, \ldots, c_d$ *belong to* P_p;
  (iv) $c_1 \kappa^m, \ldots, c_d \kappa^m$ *belong to* P_q;
where $d(i)$ *tends to infinity with* i.

From the description of κ given after the formal definition it is clear that there is a subinterval of $\{ c \mid a_i < c < a_{i+1} \}$ of length b_i on which the action of κ^m is order reversing. To obtain numbers $c_1, \ldots, c_d$ satisfying the conditions of the lemma we have only to delete from this subinterval any numbers c such that either c belongs to $\mathsf{N} \setminus P_p$ or $c\kappa^m$ belongs to $\mathsf{N} \setminus P_q$. Now κ fixes the set $\{ c \mid a_i < c < a_{i+1} \}$ setwise, and, during the construction, a number in this set is added to $\mathsf{N} \setminus P_p$ or to $\mathsf{N} \setminus P_q$ only when one is added to $\mathsf{N} \setminus P_0$. It herefore follows from Lemma 4.3 that the number of deletions we have to make is less than 2.2^i. That is, we can take $d(i)$ greater than $b_i - 2.2^i$; and the numbers b_i were chosen to make this number tend to infinity with i. The lemma follows.

LEMMA 4.6. *For distinct integers* p, q, P_p *is not boundedly enumeration reducible to* P_q.

Suppose that $P_p \leqslant_e P_q$ via U_j. We choose i so that $i > |m|$, where $m = q - p$, $i > f(p, q, j)$, and, for any PPW (k, A) elected at any time against (p, q, j) or against any triple with higher priority, a_i is greater than any element of A. (These conditions only impose lower bounds on i, so that any sufficiently large number will do.) The largest number c_d in the set $c_1, \ldots, c_d$ guaranteed by Lemma 4.5 belongs to P_p, and $P_p \leqslant_e P_q$ via U_j, so there is a pair (c_d, D) in U_j with $D \subset P_q$. We focus attention on a point in the construction when the machine M has already made the statement '(c_d, D) belongs to U_j', when all elections of PPW's against (p, q, j) or against triples with higher priorities have already been made, and at which we are again required, if possible, to ensure the existence of a PPW against (p, q, j). There cannot exist a PPW against (p, q, j) at this point, for if there were one it would remain to the end of the construction, since there are no further elections of PPW's of higher priority, and would witness against the fact that $P_p \leqslant_e P_q$ via U_j. Nor can there be a PW, for we would have elected it PPW, whereas all such elections are in the past. So we must have considered the possibility of creating and electing a PPW. In particular M has already made the statement '(c_d, D) is in U_j', and c_d, being an element of P_p, cannot have been known at this or any other time to be in $\mathsf{N} \setminus P_p$, nor similarly any element of D to be in $\mathsf{N} \setminus P_q$; so we must have considered definitions which add c_d to $\mathsf{N} \setminus P_p$. Now c_α, $\alpha = 1, \ldots, d-1$, belongs to P_p, and so, at this and every point in the construction is, like c_d, the smallest element in its MPC for $\kappa^{-p} \rho \kappa^p$. A definition could certainly be made which would fuse these MPC's, and since $c_\alpha < c_d$, it would add c_d to $\mathsf{N} \setminus P_p$. It would, of course, add also an element in the interval $\{ c \mid a_i < c < a_{i+1} \}$ to each $\mathsf{N} \setminus P_r$, and we chose i so large that this implies that it could not fall foul of either of the side conditions. The only reason that remains to explain why we did not make this definition is that it would have added an element of D to $\mathsf{N} \setminus P_q$. Since the definition fuses the MPC's for $\kappa^{-p} \rho \kappa^p$ containing c_α and c_d, it fuses the MPC's for $\kappa^{-q} \rho \kappa^q$ containing $c_\alpha \kappa^m$ and $c_d \kappa^m$. Now these numbers are in P_q, so that they are the smallest numbers in their MPC's for $\kappa^{-q} \rho \kappa^q$, and $c_d \kappa^m < c_\alpha \kappa^m$. Thus the number the definition would have added to $\mathsf{N} \setminus P_q$ is $c_\alpha \kappa^m$. It follows that $c_\alpha \kappa^m$ belongs to D for $\alpha = 1, \ldots, d-1$, so that $|D| \geqslant d-1$. But i can be taken arbitrarily large and $d = d(i)$ tends to infinity with i, so that U_j is not bounded. That is P_p is not enumeration reducible to P_q via U for U bounded, as required.

From the discussion preceding the theorem it follows that with the proof of Lemma 4.6 the proof of Theorem 4.1 is complete.

We end the paper with some more elementary considerations, designed to illuminate the need for more work in this area. If ρ is an element of $C_0 \cap R$ then by the *inequivalence relation* F of ρ we mean the set of pairs (m, n) in $\mathbb{N} \times \mathbb{N}$ such that m, n belong to different cycles of ρ. We use F rather than the equivalence relation $(\mathbb{N} \times \mathbb{N}) \setminus F$ because we want to compare F with the principal transversal P, and $(\mathbb{N} \times \mathbb{N}) \setminus F$ like $\mathbb{N} \setminus P$ is recursively enumerable. Of course the inequivalence relations of elements of $C_0 \cap R$ which are conjugate in R are (1–1)-equivalent.

THEOREM 4.7. *The principal transversal* P *and the inequivalence relation* F *of the element* ρ *of* $C_0 \cap R$ *satisfy* $P \leqslant_{tt} F$ *and* $F \leqslant_b P$.

Clearly n belongs to P if and only if (n, i) belongs to F for $i < n$, whence $P \leqslant_{tt} F$. On the other hand $F \leqslant_e P$ via U, where U is the set of pairs $((m, n), \{p, q\})$ such that $p \neq q$ and for some integers i, j, $p = m\rho^i$, $q = n\rho^j$. Since U is bounded $F \leqslant_{be} P$, and since F, P have recursively enumerable complements, $F \leqslant_b P$.

COROLLARY 4.8. $X \leqslant_{tt} Y$ *does not imply* $X \leqslant_b Y$.

For if it did, it would follow from Theorem 4.7 that for ρ in $C_0 \cap R$ the bounded-degree of the principal transversal of ρ is an invariant of its conjugacy class in R, contrary to Theorem 4.1.

Next let $c(P, Q)$ denote the relation between subsets P, Q of $\mathbb{N}$ that holds if and only if P and Q are the principal transversals of elements of $C_0 \cap R$ which are conjugate in R. Then Theorem 2.2 and Theorem 2.3 give Theorem 4.9, and Theorem 4.7 implies Theorem 4.10:

THEOREM 4.9. *Necessary conditions for* $c(P, Q)$ *are*
(i) *that both* P *and* Q *satisfy conditions* (a), (b) *and* (c) *of Theorem 2.3 and*
(ii) *that* $P \equiv_T Q$.

THEOREM 4.10. *In Theorem 4.9,* (ii) *can be replaced by*
(ii)$'$ *there exists a recursively enumerable set* X *such that* $P \leqslant_{tt} X$, $Q \leqslant_{tt} X$, $X \leqslant_b P$, *and* $X \leqslant_b Q$.

It seems reasonable to conjecture that Theorem 4.1 remains true if the bounded-degrees are replaced by truth-table degrees. If this is so, it would of course follow that $X \leqslant_b Y$ does not imply $X \leqslant_{tt} Y$. If, on the other hand, $X \leqslant_b Y$ does imply $X \leqslant_{tt} Y$ then Theorem 4.10 is a genuine strengthening of Theorem 4.9, since it would mean that $P \equiv_T Q$ could be replaced by $P \equiv_{tt} Q$. But it also seems reasonable to conjecture that Theorem 4.10 is in any case a genuine strengthening of Theorem 4.9. It would be desirable to have these points cleared up. It might also be reasonable to ask if there is a necessary and sufficient condition for $c(P, Q)$, of a simplicity comparable to that of the characterisation of principal transversals in Theorem 2.3.

REFERENCES

[1]   Graham Higman, 'On a certain infinite permutation group', *J. Algebra* **131** (1990), 359–369.

[2]   Clement F. Kent, 'Constructive analogues of the group of permutations of the natural numbers', *Trans. Amer. Math. Soc.* **104** (1962), 347–362.

[3]   H. Rogers, *Theory of recursive functions and effective computability* (McGraw-Hill, New York, 1967).

The Mathematical Institute
24-29 St. Giles
OXFORD OX1 3LB
England

ON INTERSECTIONS OF FINITELY GENERATED SUBGROUPS
OF FREE GROUPS

WALTER D. NEUMANN

Let U and V be non-trivial finitely generated subgroups of ranks u and v respectively in a free group F and let $N = U \cap V$, of rank n. In [N] Hanna Neumann, improving on a result of Howson [H], proved the inequality

$$n - 1 \leqslant 2(u - 1)(v - 1),$$

and asked if the factor 2 can be dropped. If one translates her approach (which is a slight modification of Howson's) to graph-theoretic terms, it easily shows that the answer is often "yes"—in fact, for most U the answer is "yes" for all V.

According to Gersten [G], the above problem has come to be known as the "Hanna Neumann Conjecture." Using ideas of immersions of graphs originating from Stallings ([St]), Gersten solved the problem in some special cases (his approach is close to the one of Howson and Hanna Neumann, but seems weaker in practice). I am grateful to Alan Reid for bringing Gersten's paper to my attention, and also to Peter Neumann for leading me to other literature. In particular, [I] gives the same graph-theoretical translation of Hanna Neumann's proof [1], and [Ni] and [Se] use similar methods to prove Burns' bound [B]:

$$n - 1 \leqslant 2(u - 1)(v - 1) - \min(u - 1, \, v - 1),$$

which is the best general bound known so far. We give a version of their proof in the final section.

Hanna Neumann's question can be strengthened to ask about the sum of rank $N - 1$ as N runs through a set of representatives of conjugacy classes of non-trivial intersections $N = y^{-1} U y \cap z^{-1} V z$, and as we shall describe, the bounds that we can give, including Burns' bound, remain the same.

The support of the NSF and of the Center for Mathematical Analysis in Canberra is gratefully acknowledged.
[1] as does D. E. Cohen, who also calls Hanna Neumann's question a conjecture, in his 1989 text, dedicated in part to Hanna Neumann's memory, "Combinatorial Group Theory: a topological approach" (London Math. Soc. Student Texts 14, 1989, Prop. 8.35, p. 294, and p. ii). I am grateful to O. Kegel for this reference.

1. HANNA NEUMANN'S PROOF

We shall write for short

$$\chi_0(H) = \max(0, \ \text{rank} \, H - 1)$$

if H is a finitely generated free group.

We can assume with no loss of generality that F is of rank 2, generated by elements a and b, say, and that $N = U \cap V$ is non-trivial. Let G denote the labelled figure eight graph

whose fundamental group is F. For any subgroup H of F, let $G(H)$ denote the covering of G with fundamental group H. The vertex set of $G(H)$ can be identified with the set $H \backslash F$ of right H-cosets, in which case the a-labelled edges are the pairs (Hg, Hga) and the b-labelled edges are the pairs (Hg, Hgb). If H is non-trivial, let $G_0(H)$ denote the *spine* of $G(H)$, that is, the minimal deformation retract of $G(H)$. (It is obtained by cutting off all maximal branches of $G(H)$, where a *branch* is a contractible subgraph of $G(H)$ which meets the rest of $G(H)$ only at one end of one edge: alternatively, it is the union of the supports of all reduced circuits of $G(H)$, a *reduced circuit* being a closed path that is not homotopic to a shorter closed path.) After choosing a base point, $G_0(H)$ has fundamental group a conjugate of H. If H has finite rank then $G_0(H)$ is finite and, moreover,

$$(1) \qquad\qquad 2\chi_0(H) = \sum_{p \in \text{vert} \, G_0(H)} \partial(p) - 2,$$

where $\partial(p)$ is the valency (number of incidences of edges) at vertex p. Indeed, as an equation for minus twice the euler characteristic of a finite graph, this equation is well known and is easily proved by induction.

The graph $G(N)$ is a mutual covering of the graphs $G(U)$ and $G(V)$. The projection maps $G(N) \to G(U)$ and $G(N) \to G(V)$ map the spine $G_0(N)$ into the spines $G_0(U)$ and $G_0(V)$ respectively. Let $\pi_U \colon G_0(N) \to G_0(U)$ and $\pi_V \colon G_0(N) \to G_0(V)$ denote these maps. Note that the map of vertex sets $\text{vert} \, G(N) \to \text{vert} \, G(U) \times \text{vert} \, G(V)$ is injective, so the same holds for $\pi = (\pi_U, \pi_V) \colon \text{vert} \, G_0(N) \to \text{vert} \, G_0(U) \times \text{vert} \, G_0(V)$. For any $p \in \text{vert} \, G_0(N)$, we clearly have

$$(2) \qquad \begin{aligned} 0 \leqslant \partial(p) - 2 &\leqslant \min\big(\partial(\pi_U(p)) - 2, \ \partial(\pi_V(p)) - 2\big) \\ &\leqslant \big(\partial(\pi_U(p)) - 2\big)\big(\partial(\pi_V(p)) - 2\big). \end{aligned}$$

Thus, by (1) and (2) and the injectivity of π,

$$
\begin{aligned}
2\chi_0(N) &= \sum_{p \in \text{vert } G_0(N)} \partial(p) - 2 \\
&\leqslant \sum_{p \in \text{vert } G_0(N)} \big(\partial(\pi_U(p)) - 2\big)\big(\partial(\pi_V(p)) - 2\big) \\
&\leqslant \sum_{(q,r) \in \text{vert } G_0(U) \times \text{vert } G_0(V)} \big(\partial(q) - 2\big)\big(\partial(r) - 2\big) \\
&= \left(\sum_{q \in \text{vert } G_0(U)} \big(\partial(q) - 2\big)\right)\left(\sum_{r \in \text{vert } G_0(V)} \big(\partial(r) - 2\big)\right) \\
&= 2\chi_0(U)\, 2\chi_0(V),
\end{aligned}
$$

which is the desired inequality.   □

2. Improving the proof, I.

Instead of just asking about $\text{rank}(U \cap V)$, one can ask about the ranks of all intersections $y^{-1}Uy \cap z^{-1}Vz$. Any such intersection is conjugate to one of the form $U \cap x^{-1}Vx$. Moreover, if y is in the double coset VxU, then $U \cap x^{-1}Vx$ and $U \cap y^{-1}Vy$ are conjugate. Thus we need only let x run through a set S of double coset representatives for $V \backslash F / U$. Let T be the subset of $x \in S$ with $U \cap x^{-1}Vx$ nontrivial. Denote

$$
\chi_F(U,V) = \sum_{x \in T} \chi_0(U \cap x^{-1}Vx).
$$

The size of T and $\chi_F(U,V)$ depend only on the conjugacy classes of U and V. Hanna Neumann's inequality can be strengthened as follows (see also [I2]).

PROPOSITION 2.1. T is finite and $\chi_F(U,V) \leqslant 2\chi_0(U)\chi_0(V)$.

PROOF. We can again assume F has rank 2 (since embedding F into a larger free group at worst increases the size of T and $\chi_F(U,V)$). So let $G(U)$ and $G(V)$ be as before. Let $G(U) \times G(V)$ denote the graph with vertex set $\text{vert } G(U) \times \text{vert } G(V)$ and with an a-labelled edge from (p,q) to (p',q') if and only if $G(U)$ and $G(V)$ have a-labelled edges from p to p' and q to q' respectively; similarly for b-labelled edges. We claim that the components of $G(U) \times G(V)$ are just the graphs $G(N)$ as N runs through the groups $U \cap x^{-1}Vx$, $x \in S$. Given this claim, if we denote by $G_0(U,V)$ the disjoint union of the $G_0(N)$ as N ranges over the intersections $U \cap x^{-1}Vx$ with $x \in T$ then $G_0(U,V)$ is the union of the spines of the non-contractible components of

$G(U) \times G(V)$, so it is a subgraph of $G_0(U) \times G_0(V)$. In particular, it is finite, so T is finite. Also, by equation (1),

$$(3) \qquad 2\chi_F(U,V) = \sum_{p \in \text{vert } G_0(U,V)} (\partial(p) - 2),$$

and applying the computation of section 1 to this proves the Proposition.

To see the claim, recall that we can identify the vertices of $G(U)$ and $G(V)$ with cosets of U and V in F. A component of $G(U) \times G(V)$ containing the vertex (Uy, Vz) contains the vertex (U, Vzy^{-1}), so every component contains a vertex of the form (U, Vx). The fundamental group of this component consists of all $z \in F$ with $Uz = U$ and $Vxz = Vx$; that is, $z \in U \cap x^{-1}Vx$. Moreover, another vertex (U, Vy) will be in the same component if and only if there is a $z \in F$ with $Uz = U$ and $Vxz = Vy$. That is, $Vxz = Vy$ for some $z \in U$, in other words, y is in the double coset VxU. $\qquad \Box$

Note that although T is finite, easy examples (e.g., $U = V = \langle a^n \rangle$) show that its size cannot be bounded in terms of $\chi_0(U)$ and $\chi_0(V)$. But probably the number of conjugacy classes of non-trivial subgroups $U \cap x^{-1}Vx$ can be so bounded. Only those of rank 1 are at issue, since those of rank $\geqslant 2$ number at most $\chi_F(U,V)$.

The above Proposition suggests the following:

QUESTION 2.2 (STRENGTHENED H. NEUMANN QUESTION).

$$\text{Is } \chi_F(U,V) \leqslant \chi_0(U)\chi_0(V)?$$

We shall say "$HN\{F; U, V\}$ holds" if this question has positive answer for $\{U, V\}$. It is easy to see that it holds if either U or V has finite index in F. In fact, in this case

$$\chi_F(U,V) = \chi_0(U)\chi_0(V)/\chi_0(F).$$

This is implied by the stronger result:

PROPOSITION 2.3. If U_1 has finite index d in U then $\chi_F(U_1,V) = d\chi_F(U,V)$.

PROOF. $G_0(U_1, V)$ is a d-fold covering of $G_0(U,V)$. $\qquad \Box$

In particular, if U_1 and V_1 have finite index in U and V respectively, then $HN\{F; U, V\}$ holds if and only if $HN\{F; U_1, V_1\}$ holds.

3. IMPROVING THE PROOF, II.

The above proof only used the valencies of vertices of the graphs; by taking account of the form of the vertices we can do better.

Only *nodes* (vertices of valency $\partial(p) \geqslant 3$) contribute in formulae (1) and (3). There are five forms that a node can take, and they can be listed and named as in the following poset.

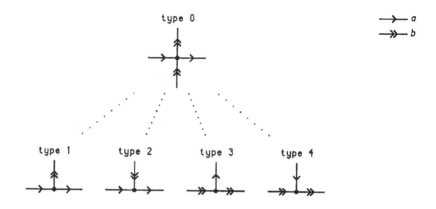

For any non-trivial subgroup H of finite rank m in F, denote the number of type i nodes of $G_0(H)$ by $k_i(H)$ for $i = 0, \ldots, 4$, so $k(H) = \sum_{i=1}^{4} k_i(H)$ is the total number of valency 3 nodes. Then equation (1) can be re-written

$$(4) \qquad 2\chi_0(H) = 2k_0(H) + k(H).$$

On the other hand, a vertex p of $G_0(U, V)$ is a node of type i only if the image vertices $\pi_U(p) \in \operatorname{vert} G_0(U)$ and $\pi_V(p) \in \operatorname{vert} G_0(V)$ are each nodes of type i or type 0. Thus, if we use equation (3) to compute $2\chi_F(U, V)$, then $p \in \operatorname{vert} G_0(U, V)$ contributes at most 2, 1, 1, or 0 according as $\pi_U(p)$ and $\pi_V(p)$ are both nodes of type 0, one of type 0 and the other of type $i \neq 0$, both of the same type $i \neq 0$, or none of the above. Thus

$$(5) \qquad 2\chi_F(U, V) \leqslant 2k_0(U)k_0(V) + k_0(U)k(V) + k_0(V)k(U) + \sum_{i=1}^{4} k_i(U)k_i(V).$$

By (4) with $H = U$, (4) with $H = V$, and (5),

$$4\chi_0(U)\chi_0(V) - 4\chi_F(U, V) \geqslant \Big(2k_0(U) + k(U)\Big)\Big(2k_0(V) + k(V)\Big)$$

$$- 2\Big(2k_0(U)k_0(V) + k_0(U)k(V) + k_0(V)k(U) + \sum_{i=1}^{4} k_i(U)k_i(V)\Big).$$

This simplifies to

$$(6) \qquad 4[\chi_0(U)\chi_0(V) - \chi_F(U, V)] \geqslant k(U)k(V) - 2\sum_{i=1}^{4} k_i(U)k_i(V)$$

$$= \sum_{i=1}^{4}(k(U) - 2k_i(U))k_i(V),$$

Suppose this is negative. Then for some i

(7) $$k(U) - 2k_i(U) < 0.$$

By symmetry, for some j

(8) $$k(V) - 2k_j(V) < 0.$$

We claim $j = i$. To see this note that the set of 4-tuples $(k_1(V), \ldots, k_4(V))$ of non-negative reals satisfying (8) for some j has four components, one for each value of j. The set of 4-tuples making (6) negative is convex, hence contained in just one of the components determined by (8), but it contains the 4-tuple $(0, \ldots, k_i(V) = k(V), \ldots, 0)$ which is in the i-th one. We have proven:

PROPOSITION 3.1. *A counter-example to $HN\{F; U, V\}$ (Question 2.2) would have to have over half the valency 3 nodes of $G_0(U)$ and over half the valency three nodes of $G_0(V)$ all of type i for some $i = 1, 2, 3,$ or 4.* □

In particular, if at most half the valency 3 nodes of $G_0(U)$ are of each type then $HN\{F; U, V\}$ holds for any V.

For "random" U the probability that over half the valency 3 nodes of $G_0(U)$ have a given type i is at most $1/16$ and is asymptotic to

$$\frac{1}{\sqrt{2\pi k(U)}} \left(\frac{3}{4}\right)^{k(U)/2}$$

as $k(U) \to \infty$. (Note that for U of fixed rank u, $k(U) = 2u - 2$ with probability 1, in the sense that among all such U with $G_0(U)$ of bounded size, the proportion with $k(U) = 2u - 2$ approaches 1 as the bound on size increases.)

4. REMARKS ON RANK 2

Suppose U has rank 2, that is, $\chi_0(U) = 1$. Then the only possibilities are

(i) $k_0(U) = 1$, $k_1(U) = \cdots = k_4(U) = 0$,

(ii) $k_0(U) = 0$, exactly two of $k_1(U), \ldots, k_4(U)$ equal 1 and the other two equal 0,

(iii) $k_0(U) = 0$, some $k_i(U)$ equals 2 and the others equal 0.

By Proposition 3.1, only in the last case could U be part of a counterexample to Question 2.2. (For example, the examples of [G; Prop. 6.12] are of type (ii).) All "small" examples of type (iii) can be changed to one of the first two types by applying an automorphism of the free group F. However, we have more than just automorphisms of F at our disposal, as we now describe.

Write $U \prec F$ if U is a finitely generated subgroup of the free group F and $HN\{F; U, V\}$ holds for all finitely generated $V < F$. The following proposition implies that the set of U of rank $\leqslant 2$ satisfying $U \prec F$ is a sub-semilattice of the semilattice of all finite rank subgroups of F.

PROPOSITION 4.1. (i) If U_1 has rank 2 and $U_2 \prec U_1 \prec F$ then $U_2 \prec F$.
(ii) If U_1 and U_2 have rank 2 then $U_1 \cap U_2$ has rank $\leqslant 2$; if, moreover, $U_1 \prec F$ and $U_2 \prec F$ then $U_1 \cap U_2 \prec F$.

LEMMA 4.2. Suppose U_1 and U_2 are subgroups of F of ranks u_1 and u_2 respectively. Suppose $U_1 \prec F$ and for each subgroup V of U_1, $HN\{G; U_2, V\}$ holds for some G containing U_1 and U_2. Then

$$\chi_F(U_1 \cap U_2, V) \leqslant \chi_0(U_1)\chi_0(U_2)\chi_0(V).$$

PROOF OF PROPOSITION. The first statement of part (ii) of the Proposition is Burns' bound quoted in the Introduction and otherwise the Proposition is immediate from the Lemma. ⬜

PROOF OF LEMMA. Let S be a set of double coset representatives for $V \backslash F / U_1$. For each x in S, choose a set of double coset representatives for $(U_1 \cap x^{-1}Vx) \backslash U_1 / (U_2 \cap U_1)$ and call it S_x, say. Since $y^{-1}U_1 y = U_1$ for each $y \in S_x$, and as S_x is part of a set of double coset representatives for $(U_1 \cap x^{-1}Vx) \backslash G / U_2$,

$$(9) \qquad \chi_G(U_2, U_1 \cap x^{-1}Vx) \geqslant \sum_{y \in S_x} \chi_0(U_2 \cap U_1 \cap y^{-1}x^{-1}Vxy).$$

Further, any double coset $Vz(U_2 \cap U_1)$ inside VxU_1 can be written in the form $Vxy(U_2 \cap U_1)$ with $y \in U_1$, and therefore also with $y \in S_x$. Thus each double coset in $V \backslash F / (U_2 \cap U_1)$ has at least one representative in $S' = \{xy \mid x \in S, y \in S_x\}$. Now, using (9) and $HN\{G; U_2, U_1 \cap x^{-1}Vx\}$ and $HN\{F; U_1, V\}$, we get

$$\chi_F(U_1 \cap U_2, V) \leqslant \sum_{xy \in S'} \chi_0(U_2 \cap U_1 \cap y^{-1}x^{-1}Vxy)$$

$$= \sum_{x \in S} \sum_{y \in S_x} \chi_0(U_2 \cap U_1 \cap y^{-1}x^{-1}Vxy)$$

$$\leqslant \sum_{x \in S} \chi_G(U_2, U_1 \cap x^{-1}Vx)$$

$$\leqslant \sum_{x \in S} \chi_0(U_2)\chi_0(U_1 \cap x^{-1}Vx)$$

$$= \chi_0(U_2)\chi_F(U_1, V)$$

$$\leqslant \chi_0(U_2)\chi_0(U_1)\chi_0(V). \qquad ⬜$$

Using automorphisms of F and Proposition 4.1 one can easily show that if U has rank 2 and $G_0(U)$ is not too large (certainly up to seven edges, but one can probably go quite a bit further) then $HN\{F; U, V\}$ holds for any V. However, there exist U of rank 2 for which $G_0(U)$ is of type (iii) whatever basis of F one chooses and U is contained in no larger rank 2 proper subgroup of F, so these techniques do not suffice to resolve the question.

The strengthened form of Burns' bound (Proposition 5.1 below) implies that $HN\{F; U, V\}$ holds if both U and V have rank 2. In particular, $U \cap x^{-1}Vx$ can have rank at most 2, and has rank 2 for at most one double coset VxU. This has some trivial but amusing consequences, whose proofs we leave to the reader.

PROPOSITION 4.3. (i) If U has rank 2 and $V \subseteq U$ then $U \cap x^{-1}Vx$ has rank $\geqslant 2$ only if $x \in U$. In particular, $HN\{F; U, V\}$ holds.
(ii) If U and V are finitely generated subgroups of a rank 2 subgroup G of the free group F then $\chi_G(U, V) = \chi_F(U, V)$.   □

5. IMPROVING THE PROOF, III.

We discuss a further strengthening of the approach of section 3.

Let G be one of the graphs $G_0(U)$, $G_0(V)$ or $G_0(U, V)$ under discussion. Rather than assigning to a node q of G one of the five types discussed in section 3, we can consider the "type" of q to be the isomorphism class of the pair $(\tilde{G}, \tilde{q})$ consisting of the universal cover $\tilde{G}$ together with a lift of the point q. Thus a "type" is an infinite contractible labelled graph with no vertices of valency 1 and with a chosen node as "base-point" (with additional properties that are not relevant to us here). Partially order such types by embeddability in each other and call two types "comparable" if they have a common lower bound in the poset of types.

Observe that the type of a node p of $G_0(U, V)$ embeds in the types of $q = \pi_U(p)$ and $r = \pi_V(p)$, so q and r are comparable. The same derivation as for equation (6) of section 3 gives

(10) $$4[\chi_0(U)\chi_0(V) - \chi_F(U, V)] \geqslant \sum_{q,r} c(q, r),$$

where the sum is over the valency 3 nodes q of $G_0(U)$ and r of $G_0(V)$ and $c(q, r) = -1$ or 1 according as q and r are comparable or not.

We describe why Burns' inequality holds in the strengthened form

PROPOSITION 5.1. $\chi_F(U, V) \leqslant 2\chi_0(U)\chi_0(V) - \min(\chi_0(U), \chi_0(V))$.

PROOF. We follow Servatius' proof [Se], which proceeds essentially as follows:

(a) Form a bipartite graph Ω with vertex set the union of the sets of valency 3 nodes of $G_0(U)$ and $G_0(V)$ and with an edge connecting a node q of $G_0(U)$ to a node

r of $G_0(V)$ if and only if q and r are comparable. An easy calculation shows that if this graph is disconnected then (10) implies the result.

(b) By embedding F in itself using the embedding with graph

all our graphs become covers of this graph, so there are no vertices of valency 4.

(c) If no vertex has valency 4 then the bipartite graph Ω of (a) is disconnected.

Nickolas [Ni] gives a simpler proof of (c), which we paraphrase. Consider a minimal counterexample to (c) (least number of vertices of $G_0(U) \cup G_0(V)$). We construct a smaller one to get a contradiction. All the nodes of $G_0(U) \cup G_0(V)$ must have the same type in the sense of section 2. By renaming we may assume it is

If a chain of two or more b's occurs in either $G_0(U)$ or $G_0(V)$, then replacing every chain of b's by a single b gives a smaller counterexample to (c), so no such chain of b's occurs. If a chain bab^{-1} occurs anywhere, then replacing each such chain by a single b gives a smaller counterexample (the graphs are still reduced since they had no chain bb), so no such chain occurs. But now a chain ba must occur somewhere, and replacing each occurrence of ba by b gives a smaller counterexample. □

Following Nickolas, we can use the same argument to get the minor result:

PROPOSITION 5.2. *Suppose* rank $U \geqslant 2$. *Then* $\chi_F(U,U) \leqslant 2\chi_0(U)^2 - 2\chi_0(U) + 1$, *and if* $V \subseteq U$ *then* $\chi_F(U,V) \leqslant 2\chi_0(U)\chi_0(V) - \chi_0(V)$.

PROOF. As above, we may assume $G_0(U)$ has only valency 3 nodes. The Servatius-Nickolas argument applied to $G_0(U)$ alone shows that these nodes can be partitioned into two non-empty mutually incomparable subsets S_1 and S_2. Equation (10) implies the first inequality, and also the second on noting that the type of any node of $G_0(V)$ is bounded above by the type of some node of $G_0(U)$, so it is incomparable with all nodes of S_1 or all nodes of S_2.

It is worth mentioning that (10) is actually an equality if $G_0(U)$ and $G_0(V)$ have no valency 4 nodes. More generally, if valency 4 nodes occur then

$$4[\chi_0(U)\chi_0(V) - \chi_F(U,V)] = \sum_{q,r} c(q,r),$$

where the sum is now over all nodes q of $G_0(U)$ and r of $G_0(V)$ and $c(q,r)$ is defined by the last column of the following table whose first column is $\{\partial(q),\ \partial(r)\}$ and whose second is the valency of the greatest lower bound type of the types of q and r (or 0 if no such type exists).

$\{3,3\}$	0	1
$\{3,3\}$	3	-1
$\{3,4\}$	0	2
$\{3,4\}$	3	0
$\{4,4\}$	0	4
$\{4,4\}$	3	2
$\{4,4\}$	4	0

It is not clear how useful this is. For instance, one might hope to answer Question 2.2 affirmatively for $\operatorname{rank} U = 2$ by showing that if $G_0(U)$ has two valency 3 nodes then each valency 3 node of $G_0(V)$ is comparable with at most one of them, but simple examples show that this can fail.

REFERENCES

[B]   Robert G. Burns, 'On the intersection of finitely generated subgroups of a free group', *Math. Z.* **119** (1971), 121–130.

[G]   S. M. Gersten, 'Intersections of finitely generated subgroups of free groups and resolutions of graphs', *Invent. Math.* **71** (1983), 567–591.

[H]   A. G. Howson, 'On the intersection of finitely generated free groups', *J. London Math. Soc.* **29** (1954), 428–434.

[I]   Wilfried Imrich, 'On finitely generated subgroups of free groups', *Arch. Math.* **28** (1977), 21–24.

[I2] Wilfried Imrich, 'Subgroup theorems and graphs', in *Combinatorial Mathematics V*, Melbourne 1976, ed. by C. H. C. Little, Lecture Notes in Math. **622**, pp. 1–27 (Springer-Verlag, Berlin Heidelberg New York, 1977).

[N]   Hanna Neumann, 'On the intersection of finitely generated free groups', *Publ. Math. Debrecen* **4** (1955–1956), 186–169. 'Addendum', *ibid.* **5** (1957–1958), p. 128.

[Ni] Peter Nickolas, 'Intersections of finitely generated free groups', *Bull. Austral. Math. Soc.* **34** (1985), 339–348.

[Se] Brigitte Servatius, 'A short proof of a theorem of Burns', *Math. Z.* **184** (1983), 133–137.

[St] John R. Stallings, 'Topology of finite graphs', *Invent. Math.* **71** (1983), 551–565.

Department of Mathematics
Ohio State University
COLUMBUS OH 43210
USA

A CHARACTERISTIC PROPERTY FOR EACH FINITE PROJECTIVE SPECIAL LINEAR GROUP

SHI WUJIE AND BI JIANXING

In [8] we have put forward the following conjecture:

CONJECTURE 1. Let G be a group and M a finite simple group. Then $G \cong M$ if and only if

(a) $\pi_e(G) = \pi_e(M)$ where $\pi_e(G)$ denotes the set of orders of elements in G, and

(b) $|G| = |M|$.

This conjecture is correct for all groups of alternating type [10], sporadic type [8], and some Lie types [9]. In this paper we prove the conjecture is correct for all $L_n(q)$, $n \geqslant 2$, using the classification of the finite simple groups.

J. G. Thompson has also put forward a related and more difficult conjecture [12]. For each finite group G, set

$$n(G) = \{ n \in \mathbf{N} \mid G \text{ has a conjugacy class } C \text{ with } |C| = n \}.$$

CONJECTURE 2. If G, M are finite groups, if $n(G) = n(M)$, and if in addition G is a nonabelian simple group while the center of M is 1, then G and M are isomorphic.

All groups G discussed will be assumed to be finite in this paper. The notation is taken from [2]; in addition, we let $\pi(G)$ denote the set of prime factors of $|G|$, and $\pi(k)$ denote the set of prime factors of a number k.

LEMMA 1. Let r be a prime and b a positive integer. Then one of the following holds.

(1) There is a prime s such that $s \mid (r^b - 1)$ but $s \nmid (r^c - 1)$ for $c < b$.

(2) $r = 2$, $b = 1$ or 6.

(3) r is a Mersenne prime and $b = 2$.

PROOF. See [14].

The project was partially supported by NSFC.

LEMMA 2. Let $a, b \in L_n(q)$, $q = p^m$, $n \geqslant 4$, $|a| = p$ and $ab = ba$. Then $\pi(|b|) \subseteq \pi(SL_{n-2}(q))$.

PROOF. If q is odd, then the conclusion is obtained by [13], §4. If q is even, then it is obtained immediately by [1], (4.3). This statement can also be checked using the matrices of $SL(n, q)$.

LEMMA 3. Let G be a finite simple group. If $p^k \,\|\, |G|$ (that is, $p^k \mid |G|$ but $p^{k+1} \nmid |G|$), p an odd prime, $k \geqslant 5$, and $|G| < p^{3k}$, then G is a group of Lie type in characteristic p.

PROOF. If G is a group of sporadic type, then $|G| > p^{3k}$ with $p^k \,\|\, |G|$, $p \neq 2$ and $k \geqslant 5$, by comparison of orders. If G is a group of alternating type then $|G| = \frac{1}{2}n!$. By the assumption, p can only be 3. Let $3^k \,\|\, \frac{1}{2}n!$; then we have $k = \sum_{j=1}^{\infty} \lfloor n/3^j \rfloor$ where $\lfloor x \rfloor$ denotes the greatest integer $\leqslant x$. From $|G| < 3^{3k}$ we infer that $n \leqslant 27$. Moreover G can only be A_6 or A_9 by a calculation. But $3^5 \nmid |A_6|$ and $3^5 \nmid |A_9|$.

Now let G be a group of Lie type in characteristic r, $r \neq p$. We derive a contradiction except when $G \cong {}^2B_2(2^{2m+1})$. If $G = L_{n+1}(r^m)$, $n \geqslant 1$, then

$$|G| = \frac{1}{d} r^{\frac{1}{2}mn(n+1)} \prod_{i=1}^{n}(r^{m(i+1)} - 1)$$

where $d = \text{g.c.d.}(n+1, r^m - 1)$. Since $p^k \,\|\, |G|$,

$$p^k \,\Big\|\, \frac{1}{d} \prod_{i=1}^{n}(r^{m(i+1)} - 1).$$

For $n \geqslant 3$,

$$\text{either} \quad p^k \mid (r^{2m} - 1) \quad \text{or} \quad p \mid \frac{1}{d} \prod_{i=2}^{n}(r^{m(i+1)} - 1).$$

If $p^k \mid (r^{2m}-1)$ then $p^k \,\|\, (r^m-1)$ or $p^k \,\|\, (r^m+1)$ as g.c.d.$(r^m-1, r^m+1) = 2$ or 1. This is contrary to $|G| < p^{3k}$. If

$$p \mid \frac{1}{d} \prod_{i=2}^{n}(r^{m(i+1)} - 1)$$

then $p \mid (r^m - 1)$ or $p \nmid (r^m - 1)$. If $p^s \,\|\, (r^m - 1)$, $s \geqslant 1$, then

$$r^{tm} + r^{(t-1)m} + \cdots + r^m + 1 \equiv t + 1 \pmod{p}$$

for $t = 1, 2, \ldots, n$. We infer that $k = sn + h$ where

$$h = \sum_{j=1}^{\infty} \lfloor (n+1)/p^j \rfloor < \frac{1}{p-1}(n+1) \leqslant \frac{1}{2}(n+1).$$

From $p^s \leqslant r^m - 1$ we get $|G| > p^{3k}$, a contradiction. If $p \nmid (r^m - 1)$ then, for a certain v, $p \mid (r^{vm} + r^{(v-1)m} + \cdots + 1)$ and $p \nmid (r^{um} + r^{(u-1)m} + \cdots + r^m + 1)$,

$u = v - 1$, $v + 1$. In this case we can easily deduce that $|G| > p^{3k}$ for $n \geqslant 3$ by the comparison of neighbouring factors of $|G|$. If $n = 2$ then

$$|G| = \tfrac{1}{d} r^{3m} (r^{2m} - 1)(r^{3m} - 1)$$

with $d = 1$ or 3. Assume $p \mid (r^m - 1)$ with $p \neq 3$. As $p \nmid (r^m + 1)$ and $p \nmid (r^{2m} + r^m + 1)$, the statement remains true. For $p = 3$, by a calculation we get $|G| > 3^{3k}$ with $3^k \parallel |G|$. For $n = 1$,

$$|G| = \tfrac{1}{d} r^m (r^{2m} - 1)$$

with $d = 1$ or 2. If $p^k \parallel |G|$ and $|G| < p^{3k}$, then $p^k = r^m + 1$ or $r^m - 1$, so $r = 2$, $p = 3$ and $k = 2$ by [3], contrary to $k \geqslant 5$. Therefore the lemma is true for $L_{n+1}(r^m)$.

Similarly we can get a contradiction when G is a group of Lie type in characteristic r, $r \neq p$, except ${}^2 B_2(2^{2m+1})$. If $G \cong {}^2 B_2(2^{2m+1})$,

$$|G| = 2^{4m+2}(2^{2(2m+1)} + 1)(2^{2m+1} - 1).$$

Since $p^k \parallel |G|$ and $|G| < p^{3k}$, $p \neq 2$, we get that $p^k \parallel (2^{2(2m+1)} + 1)$. But

$$2^{2(2m+1)} + 1 = (2^{2m+1} - 2^{m+1} + 1)(2^{2m+1} + 2^{m+1} + 1),$$

a contradiction.

LEMMA 4. Let G be a finite simple group. If $2^k \parallel |G|$, $k \geqslant 9$, and $|G| < 2^{3k}$, then G is one of the following groups:
 (i) a Lie type group in characteristic 2;
 (ii) $L_2(r)$ with r a Fermat or Mersenne prime;
 (iii) M_{24}, HS, Suz, Ru, Co_2, Co_1, Fi_{22}, B; or
 (iv) ${}^2 F_4(2)'$.

PROOF. The proof is similar to that of Lemma 3, by checking the orders of the simple groups.

LEMMA 5. If $2^k \parallel |L_n(q)|$, q odd, $n \geqslant 4$ and $k \geqslant 8$, then $2^k < q^{\frac{1}{2} n(n-2)}$.

PROOF. Since

$$|L_n(q)| = \tfrac{1}{d} q^{\frac{1}{2} n(n-1)} (q^2 - 1)(q^3 - 1) \cdots (q^n - 1)$$

where $d = \text{g.c.d.}(n, q - 1)$, $|L_n(q)| < \tfrac{1}{d} q^{(n-1)(n+1)}$. If $2^k \parallel |L_n(q)|$, then by $2 \nmid q$ and $2 \nmid (q^{2m} + q^{2m-1} + \cdots + q + 1)$, $m = 1, 2, \ldots$, we have

$$\tfrac{1}{d} 2^k q^{\frac{1}{2} n(n-1)} q^2 q^4 \cdots q^s < |L_n(q)| < \tfrac{1}{d} q^{(n-1)(n+1)}$$

where $s = n - 1$ with n odd, and $s = n - 2$ with n even. So $2^k < q^{\frac{1}{2}n(n-2)}$ for $n \geqslant 6$. When $n = 5$ we get $2^k \leqslant (q-1)^4(q+1)^2(q^2+1) < q^8$. Again by g.c.d.$(q+1, q-1) = 2$, the conclusion is true. When $n = 4$, we set $2^h \parallel (q-1)$, $h > 1$. Then $2 \parallel (q+1)$, $2 \parallel (q^2+1)$, and $k = 3h + 3$. As

$$((q+1)/2)^2((q^2+1)/2)q^4 > (q-1)^3(q+1)^2(q^2+1) = (q^2-1)(q-1)(q^4-1)$$

for $q > 8$, the conclusion holds. Since $|L_4(7)| = 2^9 \cdot 3^4 \cdot 5^2 \cdot 7^6 \cdot 19$ and $2^8 \nmid |L_4(3)|$, $2^8 \nmid |L_4(5)|$, we have the same conclusion if $q < 8$. If $2^h \parallel (q+1)$ or $2^h \parallel (q^2+1)$, $h > 1$, the discussion is the same as above. The Lemma is proved.

THEOREM 1. *Let G be a group. Then $G \cong L_2(q)$, $q = p^m$, $q > 3$, if and only if*
(a) $\pi_e(G) = \pi_e(L_2(q))$, *and*
(b) $|G| = |L_2(q)|$.

PROOF. See [9], Theorem 1.

THEOREM 2. *Let G be a group and M one of the following simple groups: $L_4(3)$, $L_5(3)$, or $L_n(2)$, $n = 3, 4, \ldots, 10$. Then $G \cong M$ if and only if*
(a) $\pi_e(G) = \pi_e(M)$, *and*
(b) $|G| = |M|$.

PROOF. It is enough to prove the sufficiency.

If $M = L_3(2)$, $L_4(2)$, $L_5(2)$ or $L_4(3)$, then $|M| < 10^8$, so the statement is true by [9], Theorem 4.

If $M = L_5(3)$, then $|G| = 2^9 \cdot 3^{10} \cdot 5 \cdot 11^2 \cdot 13$ and by the conditions G has no element of order $11 \cdot 13$. Therefore by Sylow's theorem there can be no subgroups of order $11 \cdot 13$ or $11^2 \cdot 13$ in G. By P. Hall's fundamental theorem, G is not solvable. Suppose $G = G_0 > G_1 > G_2 > \cdots > G_{k-1} > G_k = 1$ is a normal series of G where G_i/G_{i+1} is a minimal normal subgroup of G/G_{i+1}. Then there exist some i such that $\pi(G_i) \cap \{11, 13\} = \emptyset$ and $\pi(G_{i-1}) \cap \{11, 13\} \neq \emptyset$. Set $G_i = N$ and $G_{i-1} = H$, thus $G \geqslant H > N \geqslant 1$ is a normal series of G and $\overline{H} = H/N$ is a minimal normal subgroup of $\overline{G} = G/N$. Using the Frattini argument we have $11^2 \cdot 13 \mid |\overline{H}|$. Moreover $\overline{H}$ is a nonabelian simple group. Since $|\overline{H}| \leqslant |G|$ and $11^2 \cdot 13 \mid |\overline{H}|$, $\overline{H}$ can only be $L_5(3)$ by comparing the orders of simple groups and their prime-power factors (see the list of orders of simple groups in [2]). Therefore $G \cong L_5(3)$. The other cases can be treated similarly (for the orders of their elements see also [4] and [5], Table III). The theorem is proved.

THEOREM 3. Let G be a group. Then $G \cong L_3(q)$, $q = p^m$, if and only if

(a) $\pi_e(G) = \pi_e(L_3(q))$, and

(b) $|G| = |L_3(q)|$.

PROOF. 1. The case $q \equiv -1 \pmod 4$ is proved by [9], Theorem 2.

2. The case $q \equiv 1 \pmod 4$.

Suppose $M = L_3(q)$, $q \equiv 1 \pmod 4$, then $|G| = \frac{1}{d} q^3 (q^2 - 1)(q^3 - 1)$, and $\pi_1 = \pi(q(q^2 - 1))$, $\pi_2 = \pi(\frac{1}{d}(q^2 + q + 1))$ where $d = 1$ or 3, in the sense of [13], Table Ib. If G is solvable, we derive a contradiction by considering a π-Hall subgroup C of G for $\pi = \{2\} \cup \pi_2$. Let K be a minimal normal subgroup of C. If $|K|$ is odd, then the Sylow 2-subgroup S_2 of C, that is a Sylow 2-subgroup of G, acts fixed point free on K. Therefore S_2 is a cyclic or generalized quaternion group. This is impossible. So K is a 2-group and $|R| \mid (|K| - 1)$ ([9], Lemma 2), whenever R is a Sylow r-subgroup of G with $r \in \pi_2$. It follows that $\frac{1}{d}(q^2 + q + 1) \mid (|K| - 1)$, and so $\frac{1}{d}(q^2 + q + 1) \leqslant 2(q - 1)^2 - 1$ with $q - 1 = 2^k$, and $\frac{1}{d}(q^2 + q + 1) \leqslant \frac{2}{9}(q - 1)^2 - 1$ with $q - 1 \neq 2^k$. The second inequality is obviously not true. Since $\frac{1}{d}(q^2 + q + 1)$ divides $|K| - 1$, we have $\frac{1}{d}(q^2 + q + 1) = \frac{1}{w}(2(q - 1)^2 - 1)$, $w = 1, 2, 3, 4, 5$. The only solution to these equations occurs when $q = 5$. In this case $M = L_3(5)$ which has been discussed in [9], Theorem 4. Therefore G is nonsolvable. Moreover G is not a Frobenius group, by [7], Theorem 18.6. Thus G has a normal series $G \geqslant H > N \geqslant 1$ where $\pi(G/H) \subseteq \pi_1$, $\pi(N) \subseteq \pi_1$, and $\overline{H} = H/N$ is simple by [13], Theorem A. If $p \mid |N|$ then without loss of generality we can assume N is a P-group. Since G does not contain any elements of order $p_1 p$ where $p_1 \mid \frac{1}{d}(q^2 + q + 1)$ and the exponent of $p \pmod{p_1}$ is $3m$ $(q = p^m)$, the order of N is p^{3m}. Also since G has no elements of order $p_2 p$, $p_2 \mid (q + 1)$ and the exponent of $p \pmod{p_2}$ is $2m$ (q is not Mersenne, so $p_2 \nmid (q - 1)$ by Lemma 1). We infer that $p \nmid |N|$.

Now we prove $p^{3m-1} \parallel |\overline{H}|$ (that is, $p \parallel |G/H|$) when $p \mid |G/H|$. If $p^k \parallel |G/H|$, $k \geqslant 1$, then $p^k \mid |N_{\overline{G}}(\overline{P}_1)|$ as $\overline{G} = N_{\overline{G}}(\overline{P}_1)\overline{H}$ where $\overline{P}_1 \in \mathrm{Syl}_{p_1} \overline{G}$, $\overline{G} = G/N$. Since $\overline{G}$ does not contain any elements of order p^2, $p_1 p$, we deduce that $p \parallel |G/H|$ and $p^{3m-1} \parallel |\overline{H}|$.

If $p = 3$, we can assume $m \geqslant 3$ by [9], Theorem 4. In this case

$$|\overline{H}| < |G| < p^{8m} \leqslant p^{3(3m-1)}.$$

If $p \geqslant 5$, then as $\overline{H}$ is simple and $C_{\overline{G}}(\overline{H}) = 1$, we have that G/H is isomorphic to $N_{\overline{G}}(\overline{H})/\overline{H}C_{\overline{G}}(\overline{H})$ and hence to a subgroup of $\mathrm{Out}(\overline{H})$. Thus p can only divide the order m of the group of field automorphisms (see [2], Tables 1 and 5). From $m \geqslant p \geqslant 5$ we also infer that $|\overline{H}| < p^{8m} < p^{3(3m-1)}$.

When $p \nmid |G/H|$ (that is, $p^{3m} \parallel |\overline{H}|$), we have also $|\overline{H}| \leqslant |G| < p^{3(3m)}$. If $m > 1$ then $3m - 1 \geqslant 5$. Thus $\overline{H}$ is a group of Lie type in characteristic p, by Lemma 3. If

$m = 1$, then $p^3 > 2^k$ where $2^k \parallel |L_3(p)|$. It follows (by a calculation) that $\bar{H}$ is not a simple group of sporadic or alternating type. If $\bar{H}$ is of Lie type, then the characteristic of $\bar{H}$ is p by [10], Lemma 2.3. Furthermore, considering the orders of simple groups of Lie type and the orders of their maximal tori (see [5], Table III), we have $\bar{H} \cong L_3(p^m)$. For example, if $\bar{H} \cong G_2(p^{m/2})$, then $\bar{H}$ has elements of order $\frac{1}{e}(1 + p^{m/2} + p^m)$, $e = 1$ or 3. This is impossible. Hence $H \cong L_3(p^m)$ and $G \cong L_3(p^m)$. Case 2 is proved.

3. The case $p = 2$.

Let $M = L_3(2^m)$, $m \geqslant 2$. If $M = L_3(4)$, then $G \cong M$ if and only if $\pi_e(G) = \pi_e(M) = \{1,2,3,4,5,7\}$ (see [11]). If $M = L_3(8)$, then $G \cong M$ if and only if $|G| = |L_3(8)|$ and $\pi_e(G) = \pi_e(L_3(8))$ (see [9], Theorem 4). Now suppose $M = L_3(2^m)$, $m \geqslant 4$. We have $|G| = \frac{1}{d} 2^{3m}(2^{2m} - 1)(2^{3m} - 1)$ where $d = 1$ or 3. If G is solvable, we consider a $\pi(2(2^m + 1))$-Hall subgroup C of G. As in the discussion of Case 2, we infer that $\frac{1}{d}(2^m + 1) \mid (2^k - 1)$, $k \leqslant 3m$, where $2^k = |K|$ and $K \lhd C$, since G does not contain any elements of order $2r$ with $r \in \pi(2^m + 1)$ (see [1], (4.3)). As G does not contain any elements of order $2r$ with $r \in \pi(\frac{1}{d}(2^{2m} + 2^m + 1))$, similarly we get that $\frac{1}{d}(2^{2m} + 2^m + 1) \mid (2^k - 1)$, contrary to $(2^m + 1)(2^{2m} + 2^m + 1) \nmid d(2^k - 1)$, $k \leqslant 3m$, $m \geqslant 4$, and $d = 1$ or 3. Therefore G is nonsolvable, and G is not a Frobenius group. Thus G has a normal series $G \geqslant H > N \geqslant 1$ with $\bar{H} = H/N$ simple. Since G does not contain any elements of order $2p_1$ or $2p_2$ where $p_1 \mid \frac{1}{d}(2^{2m} + 2^m + 1)$ and the exponent of $2 \pmod{p_1}$ is $3m$, $p_2 \mid (2^m + 1)$ and the exponent of $2 \pmod{p_2}$ is $2m$, we can similarly infer that $2 \nmid |N|$ and $2^3 \nmid |G/H|$ because $\bar{H}$ has elements of order $\frac{1}{d}(2^{2m} + 2^m + 1)$ and has no elements of order 8. Therefore $2^t \parallel |\bar{H}|$, $t = 3m - 2$, $3m - 1$, $3m$, $m \geqslant 4$. If $t = 3m - 2$, then by $m \geqslant 4$ we have

$$|\bar{H}| \leqslant \frac{1}{2^2} 2^{3m}(2^{2m} - 1)(2^{3m} - 1) < 2^{8m-2} \leqslant 2^{9m-6} = 2^{3t}.$$

We can similarly get $|\bar{H}| < 2^{3t}$ for $t = 3m - 1$, $3m$. So $\bar{H}$ is a Lie type group of characteristic 2 by Lemma 4 and a comparison of the orders. Furthermore we have $\bar{H} \cong L_3(2^m)$ by considering the orders of simple groups of Lie type and their maximal tori. Thus $G \cong L_3(2^m)$ and the theorem is proved.

THEOREM 4. *Let G be a group. Then $G \cong L_n(q)$, $q = p^m$, $n \geqslant 2$, if and only if*

(a) $\pi_e(G) = \pi_e(L_n(q))$, *and*

(b) $|G| = |L_n(q)|$.

PROOF. It is enough to prove the sufficiency.

From Theorems 1, 2, and 3, we need only discuss the following cases: $n \geqslant 4$; $q = 2$, $n \geqslant 11$; $q = 3$, $n \geqslant 6$.

Let T_1 be the set of those prime factors p_i of $p^{mn} - 1$ for which the exponent of $p \pmod{p_i}$ is mn. Let T_2 be the set of those prime factors p_j of $p^{m(n-1)} - 1$ for which the exponent of p (mod p_j) is $m(n-1)$, and $T = T_1 \cup T_2$. Set $d = \text{g.c.d.}(p^m - 1, n)$. Then $T \cap \pi(d) = \emptyset$ by Lemma 1.

Suppose $G = G_0 > G_1 > \cdots > G_{k-1} > G_k = 1$ is a normal series of G, with G_i/G_{i+1} a minimal normal subgroup of G/G_{i+1}. Then there exists some i such that $\pi(G_i) \cap T = \emptyset$ and $\pi(G_{i-1}) \cap T \neq \emptyset$. Set $G_i = N$ and $G_{i-1} = H$; thus $G \geqslant H > N \geqslant 1$ is a normal series of G and $\bar{H} = H/N$ is a minimal normal subgroup of $\bar{G} = G/N$. Moreover $\pi(N) \cap T = \emptyset$ and $\pi(H) \cap T \neq \emptyset$.

1. $\bar{H}$ is a nonabelian simple group listed in Lemmas 3, 4, and $T \subseteq \pi(\bar{H})$.

(a) $p \nmid |N|$.

If $p \mid |N|$, then we have $1 \neq P \in \text{Syl}_p N$. By the Frattini argument $G = \text{N}_G(P)N$, thus $T \subseteq \pi(\text{N}_G(P))$. Suppose $p_s \in T_s$, $s = 1, 2$. Thus G has no elements of order $p_s p$ by Lemma 2. Therefore $|P| \equiv 1 \pmod{p_1}$ and $|P| \equiv 1 \pmod{p_2}$. If $|P| = p^t$, then $mn \mid t$ and $m(n-1) \mid t$. So $mn(n-1) \mid t$, but $p^{mn(n-1)} \nmid |L_n(q)|$, a contradiction.

(b) $\bar{H}$ is a nonabelian simple group and if $p^k \parallel |G/H|$, $k \geqslant 1$, then $k \leqslant \lceil \log_p n \rceil$ where $\lceil x \rceil$ denotes the minimal integer $\geqslant x$.

Since $\pi(H) \cap T \neq \emptyset$, we have $1 \neq P^* \in \text{Syl}_{p^*} H$ with $p^* \in \pi(H) \cap T$ and P^* cyclic. Then $\text{N}_G(P^*)$ contains elements of order p^k. Since the maximal order of p-elements contained in $L_n(q)$ is $p^{\lceil \log_p n \rceil}$, we have $k \leqslant \lceil \log_p n \rceil$. Then $k < \frac{1}{2}mn(n-1)$, and we infer that $p \mid |\bar{H}|$. Therefore $\bar{H}$ is a nonabelian simple group since $\bar{H}$ is a minimal normal subgroup and has no elements of order $p^* p$.

(c) The estimation of p^t and $|\bar{H}|$ where $p^t \parallel |\bar{H}|$.

From (b) we have $t \geqslant \frac{1}{2}mn(n-1) - \lceil \log_p n \rceil$. When $p \geqslant 5$ or $p = 3$ and $m \geqslant 2$ $(n \geqslant 4)$, or $p = 3$ and $n \geqslant 6$, we infer that

$$t \geqslant \tfrac{1}{2}mn(n-1) - \tfrac{1}{3}mn = \tfrac{1}{2}mn(n - \tfrac{5}{3}) \geqslant 5.$$

And when $p = 2$ and $n \geqslant 11$, or $p = 2$ and $m \geqslant 2$ $(n \geqslant 4)$, we infer that

$$t \geqslant \tfrac{1}{2}mn(n-1) - \tfrac{3}{8}mn = \tfrac{1}{2}mn(n - \tfrac{7}{4}) \geqslant 9.$$

The cases of $L_n(q)$ with $n \leqslant 3$, or $q = 2$ and $n \leqslant 10$, as well as $L_4(3)$ and $L_5(3)$, have already been considered. Excluding these cases we have $t \geqslant \frac{1}{2}mn(n - \frac{7}{4})$. Hence

$$|\bar{H}| \leqslant p^t \prod_{i=1}^{n-1} \left((p^m)^{i+1} - 1 \right) < p^t p^{\frac{1}{2}m(n-1)(n+2)} \leqslant p^{3t}.$$

Thus $\bar{H}$ is a simple group listed in Lemmas 3, 4.

(d) $T \cap \pi(G/H) = \emptyset$ or $p^{\frac{1}{2}mn(n-1)} \mid |\bar{H}|$.

If not, then $T_1 \cap \pi(G/H) \neq \emptyset$ or $T_2 \cap \pi(G/H) \neq \emptyset$, and $p^k \parallel |G/H|$, $k \geqslant 1$. Thus we can show $mn \mid (\frac{1}{2}mn(n-1) - k)$ or $m(n-1) \mid (\frac{1}{2}mn(n-1) - k)$ as in the proof of (a). Hence $mn \mid 2k$ or $m(n-1) \mid 2k$, so $k \geqslant \frac{1}{2}m(n-1)$. But $k \leqslant \frac{3}{8}mn$

by $k \leqslant \lceil \log_p n \rceil$ and the discussion of (c). Therefore $n = 4$ and $k = \frac{3}{2}m$. Furthermore $4m \mid (6m - \frac{3}{2}m)$ or $3m \mid (6m - \frac{3}{2}m)$ which is impossible.

(e) $T \subseteq \pi(\overline{H})$.

If $T \cap \pi(G/H) \neq \emptyset$ then $T_1 \cap \pi(G/H) \neq \emptyset$ or $T_2 \cap \pi(G/H) \neq \emptyset$. Moreover $p^{\frac{1}{2}mn(n-1)} \| |\overline{H}|$ by (d). Suppose $p_1 \in T_1 \cap \pi(G/H)$; thus $p_1 \mid (p^{mn} - 1)$ and $p_1 \nmid (p^c - 1)$ for $c < mn$. Since $p^{p_1 - 1} \equiv 1 \pmod{p_1}$, we have $mn \leqslant p_1 - 1$. Hence $p_1 \geqslant 5$. Also as $\overline{G}$ does not contain any elements of order pp_1 and $p \mid |\overline{H}|$, we see that $p_1 \notin \pi(C_{\overline{G}}(\overline{H}))$. Again as $p_1 \in \pi(G/H)$ and $\overline{G}/\overline{H}C_{\overline{G}}(\overline{H}) = N_{\overline{G}}(\overline{H})/\overline{H}C_{\overline{G}}(\overline{H})$ is isomorphic to a subgroup of $\mathrm{Out}(\overline{H})$, we infer that $p_1 \mid |\mathrm{Out}(\overline{H})|$. Since $p_1 \geqslant 5$, p_1 can only divide the order f of the group of field automorphisms, by [2], Tables 1 and 5, $mn \leqslant p_1 - 1$, and $p^{\frac{1}{2}mn(n-1)} \| |\overline{H}|$. Therefore p_1 divides $\frac{1}{2}mn(n-1)$. This is contrary to $p_1 > mn$ and p_1 prime. Using the same method we can prove $T_2 \cap \pi(G/H) = \emptyset$, so $T \subseteq \pi(\overline{H})$.

2. $G \cong L_n(q)$.

If $\overline{H}$ is a group of Lie type, then by Lemmas 3, 4 either $\overline{H}$ is a group of Lie type of characteristic p with p odd, or $\overline{H} \cong L_2(r)$ with r a Fermat or Mersenne prime and $2^t \| \frac{1}{2}(r+1)$ or $2^t \| \frac{1}{2}(r-1)$ where $2^t \| |\overline{H}|$.

(a) When $\overline{H} \cong L_{n_1}(q_1)$, $q_1 = p^{m_1}$, we have $G \cong L_n(q)$, $q = p^m$.

If $q_1 \neq 2$ or $n_1 \geqslant 8$, then by $T \subseteq \pi(\overline{H})$ and Lemma 1 we infer that $mn = m_1 n_1$ and $m(n-1) = m_1(n_1 - 1)$. Hence $m_1 = m$, $n_1 = n$, and $\overline{H} \cong L_n(q)$. The conclusion is true. If $q_1 = 2$ and $n_1 \leqslant 7$, then $p = 2$ and $mn \leqslant 7$. We have $m = 1$ and $n \leqslant 7$ (as $n \geqslant 4$). These cases have been discussed in Theorem 2.

(b) $\overline{H} \not\cong U_{n_1}(q_1)$, $q_1 = p^{m_1}$.

If not, we first assume that n_1 is even and $n_1 \geqslant 8$. Then $mn = 2m_1(n_1 - 1)$ and $m(n-1) = 2m_1(n_1 - 3)$ as $T \subseteq \pi(\overline{H})$. We get $m = 4m_1$ and $n = \frac{1}{2}(n_1 - 1)$. This is contrary to n_1 being even. If $n_1 = 6$ and $q_1 \neq 2$ we can similarly derive a contradiction using Lemma 1. If $n_1 = 6$ and $q_1 = 2$ then $mn = 10$ and $m(n-1) = 4$. This is impossible. If $n_1 = 4$ then by $mn = 2m_1(n_1 - 1)$ and $m(n-1) = m_1 n_1$ we get $n = 3$. This has been discussed in Theorem 3. When n_1 is odd and either $n_1 > 5$ or $n_1 = 5$ but $q_1 \neq 2$, we have $mn = 2m_1 n_1$ and $m(n-1) = 2m_1(n_1 - 1)$. Thus $m = 4m_1$ and $n = \frac{1}{2}n_1$. This is contrary to n_1 being odd. If $n_1 = 5$ and $q_1 = 2$, then $mn = 10$ and $m(n-1) = 4$, which is also impossible. For $n_1 = 3$ we can similarly derive a contradiction. Therefore $\overline{H} \not\cong U_{n_1}(q_1)$, $q_1 = p^{m_1}$.

(c) $\overline{H} \not\cong E_6(q_1)$, $q_1 = p^{m_1}$.

If not, we have $12m_1 = mn$ and $9m_1 = m(n-1)$. Thus $m = 3m_1$ and $n = 4$. From $p^{36m_1} \mid |\overline{H}|$ we have $p^{12m} \mid |G|$, but $12m > \frac{1}{2}mn(n-1) = 6m$, and this is contrary to $p^{\frac{1}{2}mn(n-1)} \| |G|$.

(d) $\bar{H} \not\cong L_2(q_1)$ with $q_1 = 2^r - 1$ a Mersenne prime or $q_1 = 2^{2^b} + 1$ a Fermat prime.

Assume $\bar{H} \cong L_2(q_1)$, $q_1 = 2^r - 1$, with r a prime. Then $p = 2$ and $t = r$, $2^t \parallel |\bar{H}|$. Hence $r = mn$ by $T \subseteq \pi(\bar{H})$. We have $m = 1$ and $r = n$. But $t = r = n \leqslant \frac{1}{2}n(n - 2) < t$ (see 1.(c) above), a contradiction.

If $\bar{H} \cong L_2(q_1)$, $q_1 = 2^{2^b} + 1$, then $t = 2^b$ and $p = 2$. By $T \subseteq \pi(\bar{H})$ we have $mn = 2^{b+1}$ and $m(n-1) = 2(2^b - 1)$. Thus $m = 2$, $n = 2^b$, and $t = 2^b < m(n-1) \leqslant t$, which is impossible.

Using the same method, we can prove that $\bar{H}$ is not isomorphic to $^2F_4(2)'$ or any other simple group of Lie type.

(e) $\bar{H}$ is not isomorphic to any sporadic simple group S listed in Lemma 4.

If $\bar{H} \cong S$, then $p = 2$. Suppose $S = M_{24}$; recall that $|M_{24}| = 2^{10} \cdot 3^3 \cdot 5 \cdot 7 \cdot 11 \cdot 23$. Since the exponent of 2 (mod 23) is 11 and the exponent of 2 (mod 11) is 10, $mn = 11$ and $m(n - 1) = 10$. Thus $m = 1$ and $n = 11$. But $2^{11} - 1 = 23 \cdot 89$, and $23 \cdot 89 \in \pi(L_{11}(2))$, $2 \cdot 89 \notin \pi_e(L_{11}(2))$. We easily derive a contradiction by showing that $89 \notin \pi(N)$ and $89 \notin \pi(G/H)$.

We can deal with the cases $S = HS$, Suz, Ru, Co_2, Co_1, Fi_{22}, or B in a similar fashion. The theorem is proved.

ACKNOWLEDGEMENTS

The authors want to express their profound gratitude to Professor J. G. Thompson for kind and helpful guidance and encouragement. The authors are also pleased to thank Professor W. Feit and Professor G. M. Seitz for many interesting suggestions and for sending a copy of [5], and Dr D. E. Taylor and Dr D. Parrott for their kind help.

REFERENCES

[1]   M. Aschbacher and G. M. Seitz, 'Involutions in Chevalley groups over finite fields of even order', *Nagoya Math. J.* **63** (1976), 1–91.

[2]   J. H. Conway et al., *Atlas of finite groups* (Clarendon Press, Oxford, 1985).

[3]   P. Crescenzo, 'A diophantine equation which arises in the theory of finite groups', *Adv. Math.* **17** (1975), 25–29.

[4]   U. Dempwolff and S. K. Wong, 'On finite groups whose centalizer of an involution has normal extra special and abelian subgroups, I', *J. Algebra* **45** (1977), 247–253.

[5]   W. Feit and G. M. Seitz, 'On finite rational groups and related topics' (manuscript).

[6]   B. Huppert and N. Blackburn, *Finite groups III* (Springer-Verlag, Berlin Heidelberg New York, 1982).

[7]   D. S. Passman, *Permutation groups* (Benjamin, New York, 1968).

[8]   Shi Wujie, 'A new characterization of the sporadic simple groups', in *Group Theory*, Proceedings of the Singapore Group Theory Conference held at the National University of Singapore, 1987;

ed. by Kai Nah CHENG and Yu Kiang LEONG, pp. 531–540 (de Gruyter, Berlin New York, 1989).

[9]   Shi Wujie, 'A new characterization of some simple groups of Lie type', *Contemporary Math.* **82** (1989), 171–180.

[10]   Shi Wujie and Bi Jianxing, 'A new characterization of the alternating groups', *Southeast Asian Bull. Math.* (to appear).

[11]   Shi Wujie, 'A characterization of some projective special linear groups', *J. of Math. (PRC)* **5** (1985), 191–200.

[12]   J.G. Thompson, A letter to Shi Wujie.

[13]   J.S. Williams, 'Prime graph components of finite groups', *J. Algebra* **69** (1981), 487–513.

[14]   K. Zsigmondy, 'Zur Theorie der Potenzreste', *Monatsh. Math. Phys.* **3** (1892), 265–284.

Shi Wujie   Bi Jianxing
Department of Mathematics   Department of Mathematics
Southwest-China Normal University   Liaoning University
CHONGQING   SHENYANG
China   China

ON THE CENTERS OF FREE CENTRAL EXTENSIONS
OF SOME GROUPS

V. E. SHPILRAIN

1. INTRODUCTION

Let F be a non-cyclic free group with a set $\{x_i\}_{i \in I}$ of free generators. Let ZF be its integral group ring and Δ_F the augmentation ideal, that is, the kernel of the natural homomorphism $\sigma \colon ZF \to Z$. More generally, when $1 \neq R \lhd F$ is a normal subgroup of F, we denote by Δ_R the kernel of $\sigma_R \colon ZF \to Z(F/R)$.

With every ideal $J \leq \Delta_F$ of the ring ZF we associate the normal subgroup $(J + 1) \cap F$ of F which will be denoted by $F(J)$. Each normal subgroup R of G arises in this way, as $R = F(\Delta_R)$, but in general $R = F(J)$ may hold also for ideals J different from Δ_R.

In this paper we show that the center of the group $F/[N, F]$ is precisely $N/[N, F]$ for all normal subgroups N of F which can be determined, in the way described above, by ideals of certain kinds in the ring ZF. Similar issues have been studied by several authors (see [1-4]). The main tool we use in our considerations is the well-known Fox calculus [5].

The arrangement of the paper is as follows. In Section 2 we recall some facts concerning the free differential calculus and present some technical lemmas. In Section 3 we state and prove our theorem and some of its corollaries.

2. FOX CALCULUS

In [5], Fox gave a detailed account of the differential calculus in a free group ring. He introduced free derivations as the mappings $d_i \colon ZF \to ZF$, $i \in I$, satisfying the following conditions whenever $m, n \in Z$, $u, v \in ZF$:

(1) $d_i(x_j) = \delta_{ij}$;
(2) $d_i(nu + mv) = nd_i(u) + md_i(v)$;
(3) $d_i(uv) = d_i(u)v^\sigma + ud_i(v)$.

Fox proved that these derivations have another nature as well. It turned out that the ideal Δ_F is a free left ZF-module with free basis $\{(x_i - 1)\}_{i \in I}$, and the mappings d_i are the projections to the corresponding free cyclic direct summands. Thus any element $u \in \Delta_F$ can be uniquely written in the form $u = \sum_{i \in I} d_i(u)(x_i - 1)$. (Note that on the right-hand side we always have a finite sum). That is the reason why we call $d_i(u)$ a left Fox derivative of the element u. One can define the right Fox derivations $\hat{d}_i$, $i \in I$, in a similar way, using the fact that Δ_F is also a free right ZF-module. So we have $u = \sum_{i \in I}(x_i - 1)\hat{d}_i(u)$ for any $u \in \Delta_F$. These right derivations satisfy the same conditions (1) and (2) as the left ones and a slightly different variant of condition (3):

$$\hat{d}_i(uv) = \hat{d}_i(u)v + u^\sigma \hat{d}_i(v).$$

It is of course an obvious consequence of the definitions that $d_i(1) = \hat{d}_i(1) = 0$ for all $i \in I$. The following two lemmas which are due to Fox [5] (although he considered only left derivations) will also be used throughout Section 3 without reference.

LEMMA 1. *Let* J *be an arbitrary ideal of* ZF *and let* $u \in \Delta_F$. *Then* $u \in J\Delta_F$ *(or* $u \in \Delta_F J$*) if and only if* $d_i(u) \in J$ *(or* $\hat{d}_i(u) \in J$*) for each* $i \in I$.

LEMMA 2. *Let* $y_1, y_2 \in F$ *and* $[y_1, y_2] = y_1^{-1}y_2^{-1}y_1 y_2$. *Then*

$$\hat{d}_i([y_1, y_2]) = \hat{d}_i(y_1)(y_2 - [y_1, y_2]) + \hat{d}_i(y_2)(1 - y_1^{y_2}).$$

Finally, we prove the following

LEMMA 3. *Let* $y_1, y_2 \in F$ *and let* $y_1 - 1 \in J_1$, $y_2 - 1 \in J_2$ *for some ideals* J_1, J_2 *of the ring* ZF. *Then* $[y_1, y_2] - 1 \equiv (y_1 - 1)(y_2 - 1) - (y_2 - 1)(y_1 - 1) \pmod{\Delta_F(J_1 \cap J_2)}$.

PROOF. We have $(y_1 - 1)(y_2 - 1) - (y_2 - 1)(y_1 - 1) = y_1 y_2 - y_2 y_1 = y_2 y_1([y_1, y_2] - 1)$. Hence $(y_1 - 1)(y_2 - 1) - (y_2 - 1)(y_1 - 1) - ([y_1, y_2] - 1) = (y_2 y_1 - 1)([y_1, y_2] - 1) \in \Delta_F(J_1 \cap J_2)$, as $[y_1, y_2] - 1 \in J_1 \cap J_2$.

3. THE CENTER OF $F/[N, F]$

THEOREM. *Let* $N = F(\Delta_F J)$ *for some ideal* J *of the ring* ZF, *and suppose that* $N \le F(J\Delta_F)$. *Then the center of the group* $F/[N, F]$ *is precisely* $N/[N, F]$.

PROOF. First we prove that $[N, F] \le F(\Delta_F J \Delta_F)$. It is sufficient to show that $[v, y] \in F(\Delta_F J \Delta_F)$ for any $v \in N$ and $y \in F$. Using Lemma 3 we have $[v, y] - 1 \equiv (v - 1)(y - 1) - (y - 1)(v - 1) \pmod{\Delta_F J \Delta_F}$ because $v - 1 \in J\Delta_F$. Now $v - 1 \in J\Delta_F \cap \Delta_F J$ and $y - 1 \in \Delta_F$. Hence $[v, y] - 1 \in \Delta_F J \Delta_F$.

Now let $c[N, F]$ be a central element of the group $F/[N, F]$. As $N = F(\Delta_F J)$, in order to prove that $c \in N$ it will be sufficient to show that $\hat{d}_i(c) \in J$ for each $i \in I$. Fix an arbitrary $i \in I$ and take a free generator x_j with $j \ne i$. Since c is central modulo $[N, F]$, we have $[c, x_j] \in [N, F]$ which, as we have seen, implies that $[c, x_j] - 1 \in \Delta_F J \Delta_F$. Therefore $\hat{d}_i([c, x_j]) \in J \Delta_F$. Now

$$\hat{d}_i([c, x_j]) = \hat{d}_i(c)(x_j - [c, x_j]) \equiv \hat{d}_i(c)(x_j - 1) \pmod{J \Delta_F}$$

since $[c, x_j] \equiv 1 \pmod{[N, F]}$. Hence we have that $\hat{d}_i(c)(x_j - 1) \in J \Delta_F$, and therefore $d_k(\hat{d}_i([c, x_j])) \in J$ for all $k \in I$. In particular, $d_j(\hat{d}_i([c, x_j])) = \hat{d}_i(c) \in J$. This completes the proof of the theorem.

Let $\gamma_n(G)$ denote the nth term of the lower central series of a group G. Instead of $\gamma_2(G)$ we usually write G'.

COROLLARY 1 (cf. [4]). *Let $R \lhd F$ and $n \geq 2$. Then the center of $F/[\gamma_n(R), F]$ is precisely $\gamma_n(R)/[\gamma_n(R), F]$.*

PROOF. As $\gamma_n(R) = F(\Delta_F \Delta_R^{n-1}) = F(\Delta_R^{n-1} \Delta_F)$ (see [6], §4.1), we can apply the theorem with $J = \Delta_R^{n-1}$.

COROLLARY 2. *Let $R \lhd F$, $n \geq 2$, and suppose that $F/[\gamma_n(R), F]$ is torsion-free. Then the center of $F/[\gamma_n(R), F, F, F]$ is precisely $[\gamma_n(R), F, F]/[\gamma_n(R), F, F, F]$.*

PROOF. We can apply the theorem with $J = \Delta_F \Delta_R^{n-1} \Delta_F$, because under these conditions $[\gamma_n(R), F, F] = F(\Delta_F \Delta_R^{n-1} \Delta_F^2) = F(\Delta_F^2 \Delta_R^{n-1} \Delta_F)$ follows from Theorem 2 of [7].

In particular, as $F/[\gamma_p(F'), F]$ is torsion-free for all primes p when F has rank 2 or 3 (see [8] for $p = 2$ and [9] for the general case), we have

COROLLARY 3. *If F has rank 2 or 3 and p is a prime, then the center of $F/[\gamma_p(F'), F, F, F]$ is precisely $[\gamma_p(F'), F, F]/\gamma_p(F'), F, F, F]$.*

COROLLARY 4. *For any $c \geq 2$ and $n \geq 2$, the center of $F/[(\gamma_c(F))'\gamma_n(F), F]$ is precisely $(\gamma_c(F))'\gamma_n(F))/[(\gamma_c(F))'\gamma_n(F), F]$.*

PROOF. Here we put $J = \Delta_R + \Delta_F^{n-1}$ and use a result of [10] which states that $R'\gamma_n(F) = F(\Delta_R \Delta_F + \Delta_F^n) = F(\Delta_F \Delta_R + \Delta_F^n)$ for $R = \gamma_c(F)$, $c \geq 2$, $n \geq 2$.

COROLLARY 5. *Let $R \lhd F$ such that $R \leq F'$, let $n \geq 2$, and suppose that $F/[\gamma_n(R), F]$ is torsion-free. Then the center of $F/[\gamma_n(R), F, F]$ is precisely $[\gamma_n(R), F]/[\gamma_n(R), F, F]$.*

PROOF. Theorem 3 of Stöhr [11] gives here that $[\gamma_n(R), F] = F(\Delta_F \Delta_R^{n-1} \Delta_F)$; moreover, $\gamma_n(R) = F(\Delta_R^{n-1} \Delta_F^2)$ because $R \leq F'$ implies that $\Delta_R \leq \Delta_F^2$. Hence $[\gamma_n(R), F] \leq F(\Delta_R^{n-1} \Delta_F^2)$, and we can apply the theorem with $J = \Delta_R^{n-1} \Delta_F$.

In particular, we have

COROLLARY 6. *If F has rank 2 or 3 and p is a prime, then the center of $F/[\gamma_p(F'), F, F]$ is precisely $[\gamma_p(F'), F]/[\gamma_p(F'), F, F]$.*

One could add further corollaries by using group-theoretical characterizations of other ideals, but it seems reasonable to stop here and make one concluding remark.

REMARK. On replacing the ring ZF with the ring Z_pF for an arbitrary prime p, our theorem remains valid because the augmentation ideal of Z_pF is a free left or right Z_pF-module with free basis $\{(x_i - 1)\}_{i \in I}$ (see for example [12]), which makes it possible to apply to this case all necessary results concerning the free differential calculus. But the corresponding corollaries have rather complicated form, so we have decided not to elaborate them here.

REFERENCES

[1]   J. Cossey, 'On decomposable varieties of groups', *J. Austral. Math. Soc.* **11** (1970), 340–342.

[2]   Narain Gupta, 'Certain commutator subgroups of free groups', *Comm. Pure Appl. Math.* **26** (1973), 699–702.

[3]   Narain Gupta and Frank Levin, 'Generating groups of certain soluble varieties', *J. Austral. Math. Soc.* **17** (1974), 222–233.

[4]   C. K. Gupta and N. D. Gupta, 'Generalized Magnus embeddings and some applications', *Math. Z.* **160** (1978), 75–87.

[5]   R. H. Fox, 'Free differential calculus. I. Derivation in the free group ring', *Ann. Math.* (2) **57** (1953), 547–560.

[6]   K. W. Gruenberg, *Cohomological topics in group theory*, Lecture Notes in Math. **143** (Springer-Verlag, Berlin, 1970).

[7]   V. E. Shpilrain, 'On the subgroups induced by certain ideals of a free group ring', *Vestnik Moskov. Univ. Ser. I. Mat. Mekh.* (1989), no. 6, 60–63 (in Russian).

[8]   C. K. Gupta, 'A faithful matrix representation for certain centre-by-metabelian groups', *J. Austral. Math. Soc.* **10** (1969), 451–464.

[9]   Ralph Stöhr, 'On torsion in free central extensions of some torsion-free groups', *J. Pure Appl. Algebra* **46** (1987), 249–289.

[10] C. K. Gupta, N. D. Gupta and F. Levin, 'On dimension subgroups relative to certain product ideals', in *Group Theory*, ed. by O. H. Kegel, F. Menegazzo, G. Zacher, Lecture Notes in Math. **1281**, pp. 31–35 (Springer-Verlag, Berlin, 1987).

[11] Ralph Stöhr, 'On Gupta representations of central extensions', *Math. Z.* **187** (1984), 259–267.

[12] Jacques Lewin, 'Free modules over free algebras and free group algebras: the Schreier technique', *Trans. Amer. Math. Soc.* **145** (1969), 455–465.

Department of Mathematics and Mechanics
Moscow State University
Leninskie Gory
MOSCOW 119899
USSR

GROUPS OF GENUS ZERO AND CERTAIN RATIONAL FUNCTIONS

JOHN G. THOMPSON

1. INTRODUCTION

If z is a non constant rational function of t with complex coefficients, then $C(t)$ is an extension field of $C(z)$ of degree n, where $z = \frac{f(t)}{g(t)}$, f, $g \in C[t]$, and $\max(\deg f, \deg g) = n$. If E is a splitting field for $g(T)z - f(T)$ which contains t, the Galois group G of $E/C(z)$ acts faithfully as a permutation group of the roots of $g(T)z - f(T)$ and the stabilizer of t in G is a subgroup H of index n. In [GT], R. Guralnick and I have studied the following case:

(i) $C(z)$ is a maximal subfield of $C(t)$.

(ii) The Fitting subgroup of G is $\neq 1$.

In this situation, it is the case that

$$G = NH, \quad N \cap H = 1, \quad N \triangleleft G,$$

and $|N| = n = p^e$, where p is a prime and e is a positive integer. Moreover, N is a minimal normal subgroup of G and H acts faithfully on N.

We proved that one of the following holds:

(a) H is cyclic of order $1, 2, 3, 4,$ or 6.

(b) $p = 11$, $e = 2$.

(c) $p = 7$, $e = 2$.

(d) $p = 5$, $e \leqslant 3$.

(e) $p = 3$, $e \leqslant 6$.

(f) $p = 2$, $e \leqslant 16$.

Apart from case (a), there are only finitely many groups G which satisfy (i) and (ii). As for case (a), the groups are easily determined.

It seems to me to be a worthwhile task to study each of the groups in (b) - (f), and to see what can be learned about the relationship between t and z. This paper begins the function-theoretic analysis of the groups G in question, an analysis which may require several years of hard work. One of the problems I would hope to settle is to determine just when $z \in Q(t)$. I begin with one part of Theorem 4.1 of [GT].

2. The Affine Case $p = 5$, $r = 4$, $d_1 = d_2 = d_3 = 2$, $d_4 = 3$.

I propose to construct a Galois extension $E = E_a$ of $C(z)$, for each $a \in C - \{0, 1\}$, such that

$$\operatorname{Gal} E/C(z) \cong G = NH,$$

where $N = F(G)$ is of order 25, H is dihedral of order 12, and the fixed field of H has genus zero. I follow the notation of [**GT**], only a small portion of which is needed here, and I introduce the numerical parameters which play a role in determining the algebraic nature of E.

As the notation $E = E_a$ suggests, an element a of $C - \{0, 1\}$ is of special importance. Choose b, $c \in C$ such that

$$b^2 = \frac{a}{a-1}, \quad c^2 = \frac{b+1}{b-1}.$$

The first equation forces $b \notin \{0, 1, -1\}$, and the second forces $c \notin \{0, 1, -1, i, -i\}$.

For brevity, set

$$k = C(z).$$

First, we work in $k[T]$, T an indeterminate over k, and note that

$$(T^2 + 1)^2 - \frac{z}{z-1}(T^2 - 1)^2$$

is irreducible. Let x be a zero of this polynomial in some extension field of k. Then

(1)
$$\left(\frac{x^2 + 1}{x^2 - 1}\right)^2 = \frac{z}{z-1},$$

and so $k(x) = C(z, x) = C(x)$.

Next, we work, in $C(x)[T]$ and note that

$$T^3 - \frac{x^2 - c^2}{c^2 x^2 - 1}$$

is irreducible. Let y be a zero of this polynomial in some extension field of $C(x)$. Set $F = C(x, y)$. Thus

(2)
$$y^3 = \frac{x^2 - c^2}{c^2 x^2 - 1}.$$

Let ω be a primitive cube root of 1 in C. Define maps σ, τ, ρ from $\{x, y\}$ to F:

$$\sigma(x) = x^{-1}, \quad \sigma(y) = y^{-1};$$
$$\tau(x) = -x, \quad \tau(y) = y;$$
$$\rho(x) = x, \quad \rho(y) = \omega y.$$

Since $x \neq 0$, $y \neq 0$ and $x \neq y$, these maps are well defined. It is an exercise to check that each of these maps extends to an automorphism of F which is the identity on C. These extensions are denoted by σ, τ, ρ respectively, and we see that σ, τ, $\sigma\tau$, and ρ are all non-trivial. It is an exercise to check that

$$\sigma^2 = \tau^2 = \rho^3 = 1, \quad \sigma\tau = \tau\sigma, \quad \sigma^{-1}\rho\sigma = \rho^{-1}, \quad \tau^{-1}\rho\tau = \rho.$$

So $V = \langle \sigma, \tau \rangle$ is a four group, $R = \langle \rho \rangle \triangleleft RV = D$, and D is a dihedral group of order 12 whose central involution is τ. By construction, $D = \mathrm{Gal}\, F/k$.

The field E to be constructed is an extension field of F, Galois over k, unramified over F. The construction of such fields is well understood, and some general remarks are in order.

If K is an extension field of k with $[K : k] < \infty$, denote by $g = g(K)$ the genus of K. For each positive integer n, set

$$K_{(n)} = \{ f \in K^\times \mid K(f^{1/n})/K \text{ is unramified. } \}.$$

Then $K_{(n)}$ is a multiplicative group and $K_{(n)} \supseteq K^{\times n}$. Set

$$J_n(K) = K_{(n)}/K^{\times n},$$

and call this group the group of n-division points of K. It is a basic fact, proved before anyone now living was born, and part of our education, that

(3)
$$J_n(K) \cong (\mathbf{Z}/n\mathbf{Z})^{2g}.$$

If S is an element or subset of $\mathrm{Aut}\, K$, K^S denotes the set of fixed points of S on K. Denote by $X(K)$ the set of valuation rings of K which contain C. We use this notation for various fields which contain k.

There is a bijection between $C \cup \{\infty\}$ and $X(k)$:

$$\infty \mapsto p_\infty(k) = \left\{ \frac{f}{g} \,\middle|\, f, g \in C[z],\ \deg f \leqslant \deg g,\ \text{and}\ g \neq 0 \right\};$$

$$u \mapsto p_u(k) = \left\{ \frac{f}{g} \,\middle|\, f, g \in C[z]\ \text{and}\ g(u) \neq 0 \right\}.$$

We use a similar bijection between $C \cup \{\infty\}$ and $C(x)$.

If $k \subseteq L \subseteq K$ is a tower of fields and $[K : k] < \infty$, there is a covering map

$$\phi_{K,L} : X(K) \to X(L), \quad P \mapsto \phi_{K,L}(P) = P \cap L.$$

Now we turn to F.

For each $P \in X(F)$, let D_P be the largest subgroup of D which fixes P. If $\phi_{F,k}(P) = p$, then $\phi_{F,k}^{-1}(p)$ is an orbit of D on $X(F)$, and so

$$|\phi^{-1}(\phi(P))| = |D : D_P|, \quad \phi = \phi_{F,k}, \quad P \in X(F).$$

We record some properties of $\phi_{\mathbf{C}(x),k}$. By (1), we see that if $u \notin \{\infty, 0, 1\}$, then $|\phi_{\mathbf{C}(x),k}^{-1}(P_u(k))| = 4$. We also check that

$$\phi_{\mathbf{C}(x),k}^{-1}(p_\infty(k)) = \{p_\infty(\mathbf{C}(x)), \ p_0(\mathbf{C}(x))\};$$
$$\phi_{\mathbf{C}(x),k}^{-1}(p_0(k)) = \{p_i(\mathbf{C}(x)), \ p_{-i}(\mathbf{C}(x))\};$$
$$\phi_{\mathbf{C}(x),k}^{-1}(p_1(k)) = \{p_1(\mathbf{C}(x)), \ p_{-1}(\mathbf{C}(x))\}.$$

Since $R \triangleleft D$, it follows that D acts on F^ρ; and we note that $F^\rho = \mathbf{C}(x)$. The stabilizers in D of $p_u(\mathbf{C}(x))$ are $\langle \rho, \tau \rangle$, $\langle \rho, \sigma\tau \rangle$, $\langle \rho, \sigma \rangle$ for $u = \infty, i, 1$, respectively; this being simply a matter of inspection.

Now we examine $\phi_{F,\mathbf{C}(x)}$. Since $F = \mathbf{C}(x,y)$ and (2) holds, we see that if $u \in \mathbf{C}$ and $u^2 \notin \{c^2, c^{-2}\}$, then

$$\phi_{F,\mathbf{C}(x)}^{-1}(p_u(\mathbf{C}(x))) = \left\{ p_u(\mathbf{C}(x))[y - \omega^i v] \ \middle|\ i = 0,1,2, \ v^3 = \frac{u^2 - c^2}{c^2 u^2 - 1} \right\}.$$

Similarly,

$$\phi_{F,\mathbf{C}(x)}^{-1}(p_\infty(\mathbf{C}(x))) = \left\{ p_\infty(\mathbf{C}(x))[y - \omega^i v] \ \middle|\ i = 0,1,2, \ v^3 = c^{-2} \right\}.$$

If $\varepsilon, \eta \in \{1, -1\}$, and $u = \varepsilon c^\eta$, then $p_{\varepsilon c^\eta}(\mathbf{C}(x))[y^\eta]$ is the only element of $X(F)$ which contains $p_{\varepsilon c^\eta}(\mathbf{C}(x))$.

The preceding discussion tells us that

$$\left\{ u \in \mathbf{C} \cup \{\infty\} \ \middle|\ |\phi_{F,k}^{-1}(p_u(k))| < |D| \right\} = \{\infty, 0, 1, a\};$$

for

$$\left(\frac{c^{-2}+1}{c^{-2}-1}\right)^2 = \left(\frac{c^2+1}{c^2-1}\right)^2 = \left(\frac{\frac{b+1}{b-1}+1}{\frac{b+1}{b-1}-1}\right)^2 = b^2 = \frac{a}{a-1},$$

so if $u \in \mathbf{C} - \{0, 1, a\}$, then $|\phi_{F,k}^{-1}(p_u(k))| = 12$. If $P \in \phi_{F,k}^{-1}(p_u(k))$, then $D_P = \langle \gamma \rangle$ where γ is D-conjugate to τ, $\sigma\tau$, σ, ρ, respectively, for $u = \infty, 0, 1, a$.

If $k \subseteq L \subseteq K$ with $[K : k] < \infty$ and n is a positive integer, we get a map from $J_n(L)$ to $J_n(K)$ induced by the inclusion map of L into K. To check that this is so, let's chase through an argument. We need to check that $L_{(n)} \subseteq K_{(n)}$. For each $P \in X(K)$, let t_P be a local variable at P, and let v_P be the valuation with $v_P(t_P) = 1$. Similarly, if $p \in X(L)$, t_p, v_p are a local variable at p and the valuation with $v_p(t_p) = 1$. To say that $f \in L_{(n)}$ is to say that $f \in L^\times$ and that $v_p(f) \equiv 0 \pmod{n}$ for all $p \in X(L)$. If

$P \in X(K)$, put $p = P \cap L = \phi_{K,L}(P)$. There is a positive integer r such that $t_P^r = t_p$ is a local variable at p. Then $v_P(f) = r \cdot v_p(f) \equiv 0 \pmod{n}$, so $f \in K_{(n)}$. So the inclusion of L in K gives an inclusion of $L_{(n)}$ in $K_{(n)}$. Since $L^{\times n} \subseteq K^{\times n}$, we get our induced homomorphism $J_n(L) \to J_n(K)$.

We need a criterion which forces this homomorphism to be an injection.

LEMMA 1. *If K/L is Galois and $[K:L]$ is prime to n, then $J_n(L) \to J_n(K)$ is an injection.*

PROOF. Suppose $f \in L_{(n)}$ and $f = g^n$ for some g and K. Set $m = [K:L]$. Then

$$ f^m = \left\{ \prod_{\gamma \in \mathrm{Gal}\, K/L} \gamma(g) \right\}^n . $$

So $f^m \in L^{\times n}$. By hypothesis, $(m,n) = 1$, and so $f \in L^{\times n}$. The injectivity of $J_n(L) \to J_n(K)$ follows.

Turning again to F, we check, by using the Riemann-Hurwitz formula, that

(4) $\qquad\qquad g(F) = 2, \quad g(F^\sigma) = 1, \quad g(F^{\sigma\tau}) = 1, \quad g(F^\tau) = 0.$

Set $\Lambda = \{\sigma, \sigma\tau\}$.

LEMMA 2. *If $\lambda \in \Lambda$, then τ acts by inversion on $J_n(F^\lambda)$ for all n.*

PROOF. Since $\lambda\tau = \tau\lambda$, it follows that τ acts on $J_n(F^\lambda)$ for all n. Suppose $f \in (F^\lambda)_{(n)}$. Set $e = f \cdot \tau(f)$. Since $\tau^2 = 1$, we have $e \in (F^\lambda)^\tau = F^{\langle\lambda,\tau\rangle}$. Since $g(F^\tau) = 0$, and $F^{\langle\lambda,\tau\rangle}$ is a subfield of F^τ, so also $g(F^{\langle\lambda,\tau\rangle}) = 0$. By (3), we get $e \in (F^{\langle\lambda,\tau\rangle})^{\times n} \subseteq (F^\lambda)^{\times n}$. Thus $\tau(f) \in f^{-1} \cdot (F^\lambda)^{\times n}$, and the lemma follows.

LEMMA 3. *If n is odd, then*

$$ F_{(n)} = (F^\sigma)_{(n)} \cdot (F^{\sigma\tau})_{(n)} \cdot F^{\times n}. $$

PROOF. Let $\lambda \in \Lambda$ and let $J_n(F^\lambda)^\cdot$ denote the image of $J_n(F^\lambda)$ in $J_n(F)$. By Lemma 1, $J_n(F^\lambda) \cong J_n(F^\lambda)^\cdot$, and by Lemma 2, τ inverts $J_n(F^\lambda)$.

Since the map $J_n(F^\lambda) \to J_n(F^\lambda)^\cdot$ is $\langle\lambda\rangle$-admissible, λ acts trivially on $J_n(F^\lambda)^\cdot$. It follows that $\langle\sigma, \sigma\tau\rangle$ acts trivially on $J_n(F^\sigma)^\cdot \cap J_n(F^{\sigma\tau})^\cdot$. Since $\tau \in \langle\sigma, \sigma\tau\rangle$, τ acts trivially on this intersection. Since n is odd and τ inverts $J_n(F^\lambda)$, while the map $J_n(F^\lambda) \to J_n(F^\lambda)^\cdot$ is $\langle\tau\rangle$-admissible, τ inverts the intersection, and so $J_n(F^\sigma)^\cdot \cap J_n(F^{\sigma\tau})^\cdot = \{1\}$. By (3) and (4), we get

(5) $\qquad\qquad J_n(F) = J_n(F^\sigma)^\cdot \times J_n(F^{\sigma\tau})^\cdot,$

and the lemma follows.

REMARK. The equality in the statement of the lemma holds for all n, but the more general result is not needed here.

We shall construct $J_5(F)$ by constructing $J_5(F^\lambda)$ and invoking (5).

We check that $F^\sigma = C(x + x^{-1}, y + y^{-1})$. Set

$$X = \frac{1}{y + y^{-1} - 2} - \frac{3c^2}{(c^2 - 1)^2} \; ;$$

$$Y = \frac{2i(x + x^{-1})(c^4 + 1 - c^2 y^3 - c^2 y^{-3})}{(c^4 - 1)(y + y^{-1} - 2)(y + y^{-1} + 1)} \; .$$

Thus, $C(X) = C(y + y^{-1})$ and $C(X, Y) = F^\sigma$. We find that

(6) $$Y^2 = 4X^3 - g_2 X - g_3,$$

where

(7) $$g_2 = \frac{108c^4}{(c^2 - 1)^4} + \frac{24c^2}{(c^2 - 1)^2}, \; g_3 = \frac{216c^6}{(c^2 - 1)^6} + \frac{72c^4}{(c^2 - 1)^4} + \frac{4c^2}{(c^2 - 1)^2} \; ,$$

so that $g_2^3 - 27g_3^2 = -2^4 \cdot 3^3 \cdot c^4 (c^2 - 1)^{-6}(c^2 + 1)^2$. The calculations leading to (6) are straightforward consequences of (2).

I hope to continue this work in a later paper.

REFERENCES

[GT] Guralnick, Robert M., and Thompson, John G., 'Finite Groups of Genus Zero', *J. Algebra* **131** (1990), 303–341.

University of Cambridge
CAMBRIDGE
England

DEPENDENCE OF LIE RELATORS FOR BURNSIDE VARIETIES

G. E. Wall

My purpose here is to give a new, and more illuminating, proof of an earlier result on dependence ([4], Theorem 1). I shall assume (i) the preliminary (and routine) reduction of Theorem 1 to the equivalent Theorem 1*, and (ii) the basic properties of the algebra R (see below). The reader is referred to [4] for (i) and to the forthcoming paper [5] for (ii).

The coefficient ring k for modules and algebras varies according to the context, but is always an integral domain of characteristic 0. The dependence of A, V, ... on k will be indicated where necessary by writing $A(k)$, $V(k)$, ... We write S_r for the symmetric group on $\{1, \ldots, r\}$, and kS_r for its group algebra.

Let A be the free associative algebra (with 1) on a countably infinite set of free generators x_1, x_2, ... In what follows, one should keep in mind the interpretation of A as the tensor algebra

$$T(V) = \bigoplus_{r=0}^{\infty} V^{\otimes r},$$

where V is the k-submodule generated by the x_i. Let E denote the algebra formed by those linear mappings

(1) $$\theta : A \to A$$

which commute with every V-preserving endomorphism of A; that is, with all (algebra) endomorphisms ε such that $\varepsilon(V) \subseteq V$. An element $\theta \in E$ can be specified either by the sequence of multilinear elements

(2) $$\theta(x_1 \cdots x_r) = \sum_{\pi \in S_r} a_\pi \, x_{\pi^{-1}1} \cdots x_{\pi^{-1}r} \qquad (r = 0, 1, \ldots)$$

or alternatively by the sequence of symmetry operators

(3) $$\sigma_r(\theta) = \sum_{\pi \in S_r} a_\pi \pi \in kS_r \qquad (r = 0, 1, \ldots).$$

The financial support of the Australian Research Council is gratefully acknowledged

The mapping σ which sends θ to

$$\sigma(\theta) = \big(\sigma_0(\theta),\, \sigma_1(\theta),\, \dots \big)$$

is an algebra isomorphism

$$E \cong \prod_{r=0}^{\infty} kS_r.$$

We use the customary notation $[u, v] = uv - vu$ for the Lie product in A. It is well known that the Lie subalgebra L of A generated by x_1, x_2, $\dots$ is freely generated by these elements.

LEMMA 1. If $\sigma_r(\theta)$ lies in the augmentation ideal of kS_r, then (2) is a sum of products $c_1 \cdots c_s$ with $c_i \in L$ and $s < r$.

PROOF. Since S_r is generated by adjacent transpositions, (2) is a linear combination of differences $y_1 \cdots y_i y_{i+1} \cdots y_r - y_1 \cdots y_{i+1} y_i \cdots y_r$ with $y_1, \dots, y_r$ being a rearrangement of $x_1, \dots, x_r$. But this difference is $y_1 \cdots [y_i, y_{i+1}] \cdots y_r$. $\qquad\square$

This is a convenient point at which to enunciate both the main result in its original form (Theorem 1) and the equivalent result that we shall actually prove (Theorem 1*). If m is a positive integer, let $\mathfrak{B}_m$ denote the Burnside variety formed by all groups in which the mth power of every element is the identity. We are concerned here with $\mathfrak{B}_q$, where q is a power of a prime p. A standard construction based on the lower central series associates with every group G a (graded) Lie ring $\Lambda(G)$. By a multilinear Lie relator for $\mathfrak{B}_q$ we shall mean a multilinear element

(4) $$h(x_1, \dots, x_r) \in L(\mathbf{Z})$$

such that $h(u_1, \dots, u_r) = 0$ whenever u_1, $\dots$, u_r are elements of the Lie ring $\Lambda(G)$ of a group $G \in \mathfrak{B}_q$.

Now, M. R. Vaughan-Lee in [3] for $q = p$, and later N. N. Repin in [1] for general q, constructed an infinite system of multilinear Lie relators

(5) $$g_r(x_1, \dots, x_r) \qquad (r = 1, 2, \dots)$$

of which every multilinear Lie relator is a consequence. In other words, every multilinear Lie relator is in the additive subgroup of $L(\mathbf{Z})$ generated by the elements $g_r(c_1, \dots, c_r)$ with $c_i \in L(\mathbf{Z})$. Our main theorem shows that the g_r are not independent.

THEOREM 1. If $r \not\equiv 1 \pmod{p - 1}$, then g_r is a consequence of its predecessors $g_1, \dots, g_{r-1}$.

Theorem 1* is a modified version of Theorem 1 over a new coefficient ring. Let Γ_r denote the associative subalgebra of A generated by the elements $g_s(c_1, \dots, c_s)$ with $c_i \in L$ and $s < r$. Let $\mathbf{Z}_{(p)}$ denote the ring formed by the rational numbers with denominators prime to p (this was denoted by $\mathbf{Q}^0$ in [4]).

THEOREM 1*. If $r \not\equiv 1 \pmod{p-1}$, then $g_r \in \Gamma_r(\mathbf{Z}_{(p)})$.

The key to the proof of Theorem 1* is to be found in the subalgebra R of E formed by the linear mappings (1) that commute with every L-preserving endomorphism of A, that is, with every endomorphism ε such that $\varepsilon(L) \subseteq L$. Let $\theta \in R$. Applying L-preserving endomorphisms to (2), we deduce that

$$(6) \qquad \theta(c_1 \cdots c_r) = \sum_{\pi \in S_r} a_\pi \, c_{\pi^{-1}1} \cdots c_{\pi^{-1}r}$$

whenever $c_1, \ldots, c_r \in L$.

LEMMA 2. The subalgebra Γ_r is mapped into itself by every L-preserving endomorphism ε of A and by every element θ of R.

PROOF. The elements $g_s(x_1, \ldots, x_s)$ $(1 \leqslant s < r)$ are in L, and Γ_r is generated by the set Δ of all their images under the L-preserving endomorphisms of A. Evidently $\varepsilon(\Delta) \subseteq \Delta$, whence $\varepsilon(\Gamma_r) \subseteq \Gamma_r$. As a module, Γ_r is generated by all products $d_1 \cdots d_t$ with $d_i \in \Delta$. Since $d_i \in L$, (6) shows that $\theta(d_1 \cdots d_t) \in \Gamma_r$, whence $\theta(\Gamma_r) \subseteq \Gamma_r$. $\square$

We now reproduce from [5] the salient facts about R and its relation to multilinear Lie relators. The elements of R will be specified by their associated sequences of multilinear elements (2). They are parametrized by the elements of the formal power series algebra $k[[x]]$.

For each $w \in k[[x]]$, we define an element ψ_w of E as follows: for all r, $\psi_w(x_1 \cdots x_r)$ is the $\{x_1, \ldots, x_r\}$-multilinear component of the formal power series

$$w\big((1+x_1)\cdots(1+x_r) - 1\big).$$

Then in fact $\psi_w \in R$, and the mapping $w \mapsto \psi_w$ is a module (but not algebra!) isomorphism of $k[[x]]$ onto R. This parametrization of R has the additional continuity property that, if

$$w = \sum_{n=1}^\infty w_n$$

in the sense of formal convergence in $k[[x]]$, then, for each $a \in A$,

$$\psi_w a = \sum_{n=1}^\infty \psi_{w_n} a,$$

there being only finitely many nonzero terms $\psi_{w_n} a$. Under these circumstances, we shall write

$$\psi_w = \sum_{n=1}^\infty \psi_{w_n}.$$

Let us now embed the infinite cyclic group $\langle X \rangle$ in $k[[x]]$ in the natural way:

$$X = 1 + x \qquad \text{(and so } X^{-1} = 1 - x + x^2 - \cdots \text{)}.$$

This embedding has a dual purpose. First, we have the law of multiplication

$$(7) \qquad \psi_{X^m}\psi_{X^n} = \psi_{X^{mn}} \qquad (m, n \in \mathbf{Z}),$$

from which it follows easily that R *is commutative.* Second, there is a close connection with the multilinear Lie relators for $\mathfrak{B}_q$, namely that

$$(8) \qquad g_r(x_1, \ldots, x_r) \equiv \psi_{X^q}(x_1 \cdots x_r) \pmod{\Gamma_r}.$$

Let us now take $k = \mathbf{Q}$. Then we may introduce a new variable z for formal power series by putting

$$(9) \qquad z = \ln X = \sum_{n=1}^{\infty} (-1)^{n+1} \frac{x^n}{n},$$

so that

$$X = \exp z = \sum_{n=0}^{\infty} \frac{z^n}{n!}.$$

Thus,

$$\psi_{X^m} = \sum_{r=0}^{\infty} \varepsilon_r m^r,$$

where

$$(10) \qquad \varepsilon_r = \psi_{z^r/r!} \qquad (r = 0, 1, \ldots).$$

Writing (7) in the explicit form

$$(7') \qquad (\sum \varepsilon_r m^r)(\sum \varepsilon_r n^r) = \sum \varepsilon_r (mn)^r$$

and comparing coefficients of $m^r n^s$ on both sides, we conclude that *the elements* (10) *are orthogonal idempotents.* (This argument can be made rigorous by acting on each fixed $a \in A$ with both sides of (7') and observing that $\psi_X = \sum \varepsilon_r$ implies that $\varepsilon_r a = 0$ for almost all r.)

LEMMA 3. *The element* $\sigma_r(\varepsilon_s)$ *is in the augmentation ideal of* $\mathbf{Q}S_r$ *whenever* $r \neq s$.

PROOF. For fixed r, the $\sigma_r(\varepsilon_s)$ are orthogonal idempotents in $\mathbf{Q}S_r$. An easy calculation gives

(11) $$\sigma_r(\varepsilon_r) = \frac{1}{r!} \sum_{\pi \in S_r} \pi,$$

from which the lemma follows. □

Leaving ε_0 out of account, let us now group the remaining ε_r together to form the new orthogonal idempotents

(12) $$\eta_i = \sum_{r=0}^{\infty} \varepsilon_{i+r(p-1)} \qquad (1 \leqslant i \leqslant p-1).$$

LEMMA 4. *The idempotents η_i are all contained in $R(\mathbf{Z}_{(p)})$.*

PROOF. If $\eta_i = \psi_{u_i}$, then we have to prove that $u_i \in \mathbf{Z}_{(p)}[[x]]$. Let $\mathbf{Q}_p$, $\mathbf{Z}_p$ denote respectively the field of p-adic numbers and the ring of p-adic integers. Since $u_i \in \mathbf{Q}[[x]]$ and $\mathbf{Q} \cap \mathbf{Z}_p = \mathbf{Z}_{(p)}$ in $\mathbf{Q}_p$, it will be sufficient to show that $u_i \in \mathbf{Z}_p[[x]]$.

Now, it is well known that the primitive $(p-1)$th root of unity in the residue class field $\mathbf{Z}_p/p\mathbf{Z}_p \cong \mathbf{Z}/p\mathbf{Z}$ can be lifted to a primitive $(p-1)$th root of unity ω in $\mathbf{Z}_p$. Since

$$\exp(\omega^i z) - 1 = \sum_{j=1}^{p-1} \omega^{ij} u_j \qquad (1 \leqslant i \leqslant p-1)$$

and the Vandermonde determinant $\Delta(\omega, \omega^2, \ldots, \omega^{p-1})$ is a unit in $\mathbf{Z}_p$, it will be sufficient to prove that $\exp(\omega^i z) \in \mathbf{Z}_p[[x]]$.

But in fact $\exp(\lambda z) \in \mathbf{Z}_p[[x]]$ whenever $\lambda \in \mathbf{Z}_p$. For

$$\exp(\lambda z) = \sum_{r=0}^{\infty} \binom{\lambda}{r} x^r;$$

and since the continuous mapping $\lambda \mapsto \binom{\lambda}{r}$ of the closed set $\mathbf{Z}_p$ into $\mathbf{Q}_p$ maps the dense subset $\{1, 2, \ldots\}$ of $\mathbf{Z}_p$ into $\mathbf{Z}_p$, it follows that $\binom{\lambda}{r} \in \mathbf{Z}_p$ for all $\lambda \in \mathbf{Z}_p$. □

All the tools needed to prove Theorem 1* have now been assembled. The proof will be carried out within $A = A(\mathbf{Z}_{(p)})$.

By Lemma 4, $\eta_1 \in R = R(\mathbf{Z}_{(p)})$. Therefore, by (8) and Lemma 2,

$$\eta_1 g_r(x_1, \ldots, x_r) \equiv \eta_1 \psi_{X^q}(x_1 \cdots x_r) \pmod{\Gamma_r}.$$

Now, an easy calculation gives $\eta_1 x_1 = x_1$. Hence, by (6), $\eta_1 c = c$ for all $c \in L$ and so, in particular, the left hand side of the equation above is just $g_r(x_1, \ldots, x_r)$. On the right hand side, $\eta_1 \psi_{X^q} = \psi_{X^q} \eta_1$ because R is commutative. Thus, we get

(∗) $$g_r(x_1, \ldots, x_r) \equiv \psi_{X^q}(\eta_1(x_1 \cdots x_r)) \pmod{\Gamma_r}.$$

Let $c_1, \ldots, c_s \in L$, where $s < r$. By (8),

$$\psi_{X^q}(x_1 \cdots x_s) \equiv g_r(x_1, \ldots, x_s) \pmod{\Gamma_s}$$

and so

(†) $\psi_{X^q}(x_1 \cdots x_s) \equiv 0 \pmod{\Gamma_r}$.

Let ε be any L-preserving endomorphism of A such that $\varepsilon(x_i) = c_i$ $(1 \leqslant i \leqslant s)$. Since ε commutes with ψ_{X^q},

$$\begin{aligned}
\psi_{X^q}(c_1 \cdots c_s) &= \psi_{X^q}(\varepsilon(x_1 \cdots x_s)) \\
&= \varepsilon \psi_{X^q}(x_1 \cdots x_s).
\end{aligned}$$

Therefore, by (†) and Lemma 2,

(∗∗) $\psi_{X^q}(c_1 \cdots c_s) \equiv 0 \pmod{\Gamma_r}$.

So far we have placed no restriction on r. Assume now that $r \not\equiv 1 \pmod{p-1}$. By (12) and Lemma 3, $\sigma_r(\eta_1)$ is in the augmentation ideal of $\mathbf{Z}_{(p)} S_r$. Hence, by Lemma 1, $\eta_1(x_1 \cdots x_r)$ is a sum of products $c_1 \cdots c_s$ with $c_i \in L$ and $s < r$. It now follows from (∗) and (∗∗) that

$$g_r(x_1, \ldots, x_r) \equiv 0 \pmod{\Gamma_r},$$

as we had to prove.   □

We conclude with some remarks about the idempotents that played a crucial part in the proof. Since $\psi_X = \sum_{r=0}^{\infty} \varepsilon_r$ is the identity element of $R = R(\mathbf{Q})$, the ε_r provide a primitive idempotent decomposition of the identity in R. The components $\varepsilon_r A$ in the corresponding vector space direct decomposition of $A = A(\mathbf{Q})$ are easily identified. By (11) and (6),

$$\varepsilon_r(c_1 \cdots c_r) = \frac{1}{r!} \sum_{\pi \in S_r} c_{\pi^{-1}1} \cdots c_{\pi^{-1}r}$$

whenever $c_1, \ldots, c_r \in L$. It follows that the subspace $L^{(r)}$ of A spanned by the rth powers of the elements of L is contained in $\varepsilon_r A$. But it is easy to see that A is the sum of the $L^{(r)}$, so that in fact

(13) $\varepsilon_r A = L^{(r)}$.

This identification indicates the close relation between the ε_r and the projections ϕ_r in L. Solomon's paper [2].

We remark finally that ε_0 and the η_i provide a *primitive* idempotent decomposition of the identity in $R(\mathbf{Z}_{(p)})$. For if η_i were not primitive then it would split into the sum of two idempotents, one of which had the form $\varepsilon_r + \varepsilon_{r'} + \varepsilon_{r''} + \cdots$, where $p \leqslant r < r' < r'' < \cdots$. But such an element has the form ψ_w, where

$$w = \frac{x^r}{r!} + \text{terms of higher degree},$$

and clearly $w \notin \mathbf{Z}_{(p)}[[x]]$.

REFERENCES

[1] N. N. Repin, 'Multilinear identities in the Lie rings associated with periodic groups', *Izv. Akad. Nauk SSSR Ser. Mat.* **52** (1988), 1091–1101; *Math. USSR—Izv.* **33** (1989), 413–423.

[2] L. Solomon, 'On the Poincaré-Birkhoff-Witt theorem', *J. Combin. Theory* **4** (1968), 363–375.

[3] M. R. Vaughan-Lee, 'The restricted Burnside problem', *Bull. London Math. Soc.* **17** (1985), 113–133.

[4] G. E. Wall, 'On the multilinear identities which hold in the Lie ring of a group of prime-power exponent', *J. Algebra* **104** (1986), 1–22.

[5] G. E. Wall, 'Multilinear Lie relators for varieties of groups' (in preparation).

Department of Pure Mathematics
UNIVERSITY OF SYDNEY, NSW 2006
Australia

Corrigenda to the paper
"On the rank of the intersection of subgroups of a Fuchsian group"

R. G. BURNS

The paper in question appeared on pp. 165–187 of *Proc. Second Internat. Conf. Theory of Groups*, Canberra 1973, ed. by M. F. Newman, Lecture Notes in Math. **372**, (Springer-Verlag, Berlin Heidelberg New York, 1974). It purported to give an upper bound, in terms of m and n alone, for the rank of the intersection of two subgroups of ranks m and n in a Fuchsian group. This is an announcement only, of errors, some serious, in that paper. The errors have been corrected and a corrected version may eventually be published; in the meantime, full details of the corrections may be obtained from the author. The latter hereby thanks Dr D. E. Cohen for pointing out some of the errors. In what follows, the page and reference numbers are as in that paper, and so is the notation.

Page 165, first line. $LF(2, \mathbf{R})$ is not, as stated, $SL(2, \mathbf{R})$, but isomorphic to $PSL(2, \mathbf{R})$.

Page 166, part (i) of Theorem 1.1 and line 15 from the bottom. The paper [16] has been corrected: the rank of a Fuchsian group is at least $n + t - 2$. Hence a version of part (i) of Theorem 1.1 holds without the condition $t = 0$.

Page 166, last line. One also needs that not both $n = 1$ and $t = 1$.

Section 3. The proof of Theorem 3.1 is fallacious, and that theorem is to be replaced by a different, less precise analogue of the 'Howson—Hanna Neumann formula'. Thus all of Section 3 from line 17 of page 170 needs to be replaced.

Section 4. Corollary 4.4 is no longer required. Corollary 4.5 is to be replaced since its proof is fallacious. The collapse of Theorem 3.1 entails that of Corollary 4.6, which therefore yields to a weaker result (which now replaces part (ii) of Theorem 1.1).

Section 5. Theorem 5.1 needs to be changed in the light of the above. In Corollary 5.3 one has in fact: (i) $\gamma = \alpha + \beta - 1$; and (ii) $\gamma = \alpha$.

Once the corrections and resulting changes have been made, the bound in the corrected version of part (iii) of Theorem 1.1 can be computed as on pages 181, 182, yielding a polynomial of degree 8 in m and n.

York University
DOWNSVIEW, Ontario M3J IP3
Canada